ROUTLEDGE LIBRARY EDITIONS:
ENVIRONMENTAL POLICY

Volume 2

# ENVIRONMENTAL POLICY AND IMPACT ASSESSMENT IN JAPAN

# ENVIRONMENTAL POLICY AND IMPACT ASSESSMENT IN JAPAN

BRENDAN F. D. BARRETT AND
RIKI THERIVEL

Routledge
Taylor & Francis Group

LONDON AND NEW YORK

First published in 1991 by Routledge

This edition first published in 2019
by Routledge
2 Park Square, Milton Park, Abingdon, Oxon OX14 4RN

and by Routledge
52 Vanderbilt Avenue, New York, NY 10017

*Routledge is an imprint of the Taylor & Francis Group, an informa business*

*British Library Cataloguing in Publication Data*
A catalogue record for this book is available from the British Library

ISBN: 978-0-367-18894-8 (Set)
ISBN: 978-0-429-27423-7 (Set) (ebk)
ISBN: 978-0-367-18910-5 (Volume 2) (hbk)
ISBN: 978-0-429-19916-5 (Volume 2) (ebk)

**Publisher's Note**
The publisher has gone to great lengths to ensure the quality of this reprint but
points out that some imperfections in the original copies may be apparent.

**Disclaimer**
The publisher has made every effort to trace copyright holders and would welcome
correspondence from those they have been unable to trace.

# Environmental Policy and Impact Assessment in Japan

Brendan F.D. Barrett and
Riki Therive

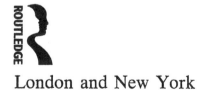

London and New York

First published 1991
by Routledge
11 New Fetter Lane, London EC4P 4EE

Simultaneously published in the USA and Canada
by Routledge
a division of Routledge, Chapman and Hall, Inc.
29 West 35th Street, New York, NY 10001

© 1991 Brendan F.D. Barrett and Riki Therivel

Disk conversion by Columns of Reading
Printed and bound in Great Britain by
Mackays of Chatham PLC, Chatham, Kent

*British Library Cataloguing in Publication Data*
Barrett, Brendan F. D. *1960–*
   Environmental Policy and Impact Assessment in Japan. –
   (The Natural Environment. Problems and Management Series,
   ISSN 0956–7488).
   1. Japan. Environment. Pollution. Control measures.
   Policies of government
   I. Title II. Therivel, Riki, *1960–* III. Series
   363.7360952

   ISBN 0–415–03852–9

*Library of Congress Cataloging in Publication Data*
Barrett, Brendan F. D., 1960–
   Environmental Policy and Impact Assessment in Japan/
   Brendan F. D. Barrett and Riki Therivel.
   p.  cm. – (Natural Environment–Problems and Management Series)
   Includes bibliographical references.
   ISBN 0–415–03852–9
   1. Environmental policy–Japan.  2. Environmental protection–
   Japan.  3. Environmental impact analysis–Japan.  I. Therivel,
   Riki, 1960–  II. Title.  III. Series.
   HC465.E5B37  1990
   363.7'058'0952–dc20                                         90–8511
                                                               CIP

To our parents
Brigitte and William Therivel and Ann and Peter Barrett

# Contents

# Figures

# Tables

# Preface

We met at Kyoto University in early 1987. We were both researchers sponsored by the Japanese Ministry of Education (Monbusho) and both interested in Japan's environmental problems, so we decided to pool our resources.

We soon realized that the literature on Japanese environmental issues falls into several broad groups. Official reports by the Japanese government or quasi-governmental organizations represent the consensus opinion of Japan's industry–government hegemony; they are well-documented and informative, but avoid contentious issues. Some internal government reports discuss contentious issues but remain confidential. Reports by environmental groups are limited by the lack of public information, and are often not translated into English. Articles or books by non-Japanese also address topics which arouse conflict, but the books are becoming outdated, and the articles by necessity deal only with specific issues.

This book is an attempt to fill our perceived need for a reasonably comprehensive work which summarizes the current state of environmental policies in Japan and points to possible problems of these policies. It tries to add a critical analysis to the government publications, to update non-Japanese materials, and to offer a broader circulation to some Japanese-language works.

The Japanese persistently claim that most outsiders do not really understand Japan. On the surface Japan appears to be monistic, but beneath that surface there are a myriad of conflicts which closely parallel those of Western societies. However, these pluralistic tendencies are limited by the fact that many explanations for social, economic and political phenomena have become established and accepted by the Japanese with an affirmation which is found in the modern Western society only when natural phenomena are discussed. Views which contradict these explanations tend to be frowned upon. In addition, objective analysis often becomes confused with criticism especially when it originates from outside of Japan.

This book is written from a Western perspective. Although we would have liked to be objective, we found it impossible to escape the influence of our Western values and biases. As such we are at times critical of a system which seems to operate on principles so different from the ones we are used to. Because this work challenges many established explanations, it has been described by some Japanese academics as over-critical and lacking in understanding of Japan's true situation.

Environmental planning and protection in the Western sense is only one method for promoting wise use of natural resources, and is based on assumptions about society and the individual which are not universally accepted. Judging all methods by the yardsticks of environmental planning – analysis and control of nature – does not do them justice, and measuring Japanese achievements by Western standards may well be misleading.

On the other hand, with environmental problems reaching global proportions, we believe that it is time to stop using national boundaries and cultural characteristics as an excuse for poor environmental management. We hope that this book provokes a reassessment of the effectiveness of Japan's environmental policies, and maybe even promotes improved environmental protection.

# Acknowledgements

The book would not have been possible without the help of many colleagues and friends. We are particularly grateful to Toshio Hase, who not only wrote Chapter 14 and introduced us to countless interesting topics and people, but also offered us his friendship and advice. We are thankful for the invaluable assistance of Professor Tsuneo Tsukatani, Maggie Suzuki, Professor Hiroshi Takatsuki, Professor Takeo Suzuki, Professor Fumio Takeda, Motokazu Iwata, Yoshiaki Shimada, Dr Tsuneyuki Morita, Professor Toru Iwama, Junichi Ishikura, Professor Jun Ui and many others who generously shared their time, knowledge and often their lives with us. We would also like to express our gratitude to Rob Woodward for the maps and the photo, to Evelyn Martin for the flow-charts and for typing Appendix C, and to Professor John Glasson for his encouragement and support. We also gratefully acknowledge the permission from Kajina Institute Publishing Company to reprint materials from *Environmental Assessment in Practice*.

Our work was partially funded by the Japanese Ministry for Education and Science and Oxford Polytechnic's Socio-Economic Impact Assessment Unit, and was supported by the laboratories of Professor Hisashi Sumitomo and Professor Takuma Takasao, and the Economic Research Institute at Kyoto University. Yoshitaka Emoto and all the staff at the Access Research Firm, Osaka were also very supportive.

We would like to express our warmest thanks to Chizu Kawahara, Tim O'Hara, Carolyn Logan and Joe Banerjee for making our lives sane and enjoyable, for help with the translation work, and for reading and commenting on various chapters. Finally, we thank each other for patience, humour, and hard work through the endless trans-continental revisions.

| | |
|---|---|
| Brendan Barrett | Riki Therivel |
| Economics Research Institute | School of Planning |
| Kyoto University | Oxford Polytechnic |

# Abbreviations

| | |
|---|---|
| *ADC* | Aviation Deliberation Committee (MoT) |
| *APA* | agricultural promotion area |
| *art* | article |
| *bn* | Billion (US) |
| *BOD* | biochemical oxygen demand |
| *CCEPC* | Central Council on Environmental Pollution Control (EA) |
| *CCZ* | city consolidation zone |
| *CDZ* | city development zone |
| *CNDP* | comprehensive national development plan |
| *CO* | carbon monoxide |
| *COD* | chemical oxygen demand |
| *CPA* | city planning area |
| *EA* | Environment Agency |
| *EC* | European Community |
| *EIA* | environmental impact assessment |
| *EIS* | environmental impact statement |
| *EPA* | Economic Planning Agency |
| *EQS* | environmental quality standard |
| *EUZ* | existing urban zone |
| *GNP* | gross national product |
| *ha* | hectare |
| *HSB* | Honshu–Shikoku Bridge project |
| *IA* | Inducement Area |
| *IUCN* | International Union for the Conservation of Nature |
| *JHPC* | Japan Highway Public Corporation |
| *KIA* | Kansai International Airport |
| *LDP* | Liberal Democratic Party |
| *m* | Million |
| *MAFF* | Ministry of Agriculture, Forestry and Fisheries |
| *MHA* | Ministry of Home Affairs |
| *MHW* | Ministry of Health and Welfare |
| *MITI* | Ministry of International Trade and Industry |

| | |
|---|---|
| *MoC* | Ministry of Construction |
| *MoT* | Ministry of Transport |
| *MPN* | Most probable number |
| *NCR* | National Capital Region |
| *NEPA* | National Environmental Policy Act (US) |
| *NIA* | New Ishigaki Airport |
| *NLA* | National Land Agency |
| *NLUPA* | National Land Use Planning Act |
| *NO$_2$* | nitrogen dioxide |
| *NOx* | nitrogen oxide |
| *OPEC* | Organization of Petroleum Exporting Countries |
| *PCB* | Polychlorinated Biphenyl |
| *PM* | prime minister |
| *ppm* | parts per million |
| *PPP* | Polluter Pays Principle |
| *PWARL* | Public Water Areas Reclamation Law |
| *REMP* | regional environmental management plan |
| *RPA* | Relocation Promotion Area |
| *RPCP* | regional pollution control programmes |
| *tr* | Trillion (US) |
| *SCZ* | suburban consolidation zone |
| *SO$_2$* | sulphur dioxide |
| *SOx* | sulphur oxide |
| *SS* | suspended solids |
| *TBH* | Trans-Tokyo Bay Highway |
| *TMR* | Tokyo Metropolitan Region |
| *UCA* | urbanization control area |
| *UK* | United Kingdom |
| *UPA* | urbanization promotion area |
| *US* | United States of America |
| *WECPNL* | weighted equivalent continuous predicted noise level |
| *WWFN* | World Wide Fund for Nature |

**Note**

Billions used in the text are US billions (i.e. 1,000 × million). Trillions used in the text are US trillions (i.e. 1,000 × billion).

**Figure I.1 Prefectures**

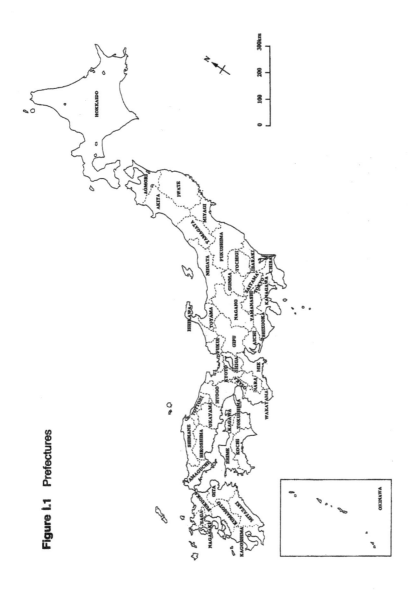

**Figure I.2** Major cities and relevant towns

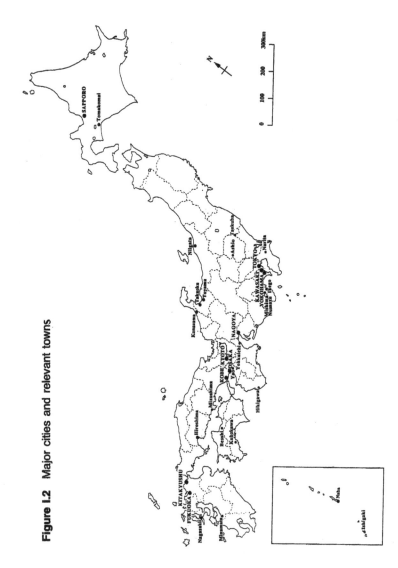

Note: Capitalized cities are designated cities.

**Figure I.3**  Regions and water bodies

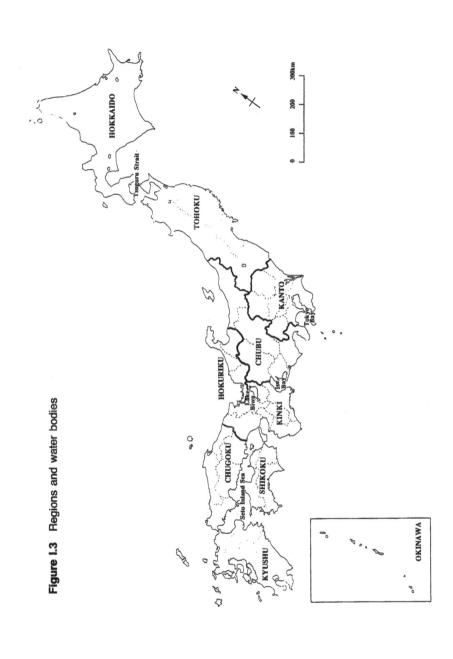

# Chapter one

# Introduction

Japan is a leader in the development of pollution control technology and one of the world's major polluters. It has some of the world's strictest environmental quality standards and some of its most environmentally damaged areas. The Japanese are known as a nature-loving people but at the same time they are responsible for widespread environmental destruction. The administration is keen to become a leader in solving global environment problems yet causes international tension over such environmental issues as whaling and large developments overseas.

Japan is a relatively small country with limited resources. In the last 100 years it has transformed itself from a predominantly agricultural country into one of the biggest centres of industrial production. The growth of major manufacturing industries, particularly after the Second World War, brought about the modernization of the Japanese archipelago, but with it also the concentration of people and industries in the cities and extensive pollution problems. Public outcry over environmental pollution in the 1960s forced the government to embark on a major offensive against pollution. In the 1970s environmental policies were developed and strict environmental standards were enforced. During the 1980s, however, environmental policy declined as a result of rising economic concerns and widespread public apathy. As we enter the 1990s, the future of Japan's environmental policy is torn between the requirements of a healthy economy and a healthy environment. Increasing awareness of global environmental problems could provide the impetus to improve on the poor performance of the 1980s but this could be counter-balanced by the desire to make ever greater economic gains. This scenario has its parallels in many other countries.

Japan differs from other countries, however, in the magnitude and polarity of these changes. After the Second World War all of the nation's efforts were put into developing its economic base. Environmental problems had to reach a critical state, with hundreds of people sick or dying of pollution-related illnesses, before their existence was acknowledged. In response the government instituted some of the world's strictest anti-pollution measures. Now, with equal energy, Japan is once again developing, safe in the expectancy of having solved its environmental problems.

How Japan balances economic growth and environmental protection has far-reaching consequences both within and outside the country. For other highly industrialized countries Japan is an alternate scenario: the great historical and cultural differences between Japan and these countries have caused them to differ not only in their views of technology and development, but in how they deal with their environmental impacts. For rapidly developing countries such as Taiwan and Korea, Japan is both an inspiration and a warning: its undeniable economic success has been achieved at great social and environmental cost, and hopefully this knowledge will cause these countries to consider carefully their balance of economic and environmental priorities.

This book examines the development and current state of environmental policy, and in particular of environmental impact assessment, in Japan. The book is in three parts: environmental policy, environmental impact assessment and case studies. The first part reviews interest group participation in environmental policy-making, the history and current state of environmental problems in Japan, the land use and economic planning framework, and Japan's current environmental policies. The second part discusses the development of and procedures for EIA in Japan. The third part addresses five development projects which illustrate the conflict between economic growth and environmental protection in Japan, and the application of EIA procedures.

Although most readers will be familiar with EIA, we provide a brief definition here for those for whom it is a new concept. Environmental impact assessment is the systematic examination of the effects of a project or policy on the environment. EIA is expected to provide policy-makers with environmental information under the assumption that the more information is considered, the better the subsequent decision. In particular, EIA is meant to be a preventive measure used in a project's planning stage to avoid environmentally harmful development; this contrasts with other environmental policies which are primarily *post-facto* reactive measures. EIA is one part of a more comprehensive process aimed at improving resource allocation and environmental management.

The end-product of an EIA process is a combination of text and diagrams, an Environmental Impact Statement. Generally an EIS includes:

— a description of the proposed project;
— a description of environmental conditions at the proposed site;
— a description and comparison of alternatives to the proposed project and site;

— a description of environmental factors likely to be affected by the proposed project;
— a discussion of possible methods for mitigating these effects; and
— an evaluation of the effects.[1]

EIA has been an integral part of the planning process in the US for almost twenty years.[2] It is also currently of great interest in Europe, since the European Community Directive on EIA took effect in July 1988. Other countries such as Australia, Canada and China have implemented EIA regulations or guidelines, and still others are considering doing so.[3] In Japan in 1984, after eight years of negotiations and revisions, the Environment Agency abandoned its efforts to introduce an EIA legislation,[4] and instead the Cabinet passed a (non-mandatory/non-enforceable) decision which incorporates EIA procedures.

EIA has been described as a Western elitist solution to a perceived need for environmental protection.[5] The socio-economic, geographical and political realities of a given nation will determine the degree to which it will adopt such concepts as EIA. This book examines the process by which EIA was adopted in Japan, and what its influence has been. It argues that EIA in Japan became 'Japanized' during the process of adoption, and proposes that EIA in Japan has lost its original function – environmental protection – and instead is used primarily to coordinate various agencies' environmental activities.

Some Japanese academics have described environmental impact assessment as environmental awasement. The verb *awaseru* means 'to match, bring into line'; environmental awasement means that a development's impacts are brought into line with environmental standards. This is indeed a misinterpretation of the concept of EIA in that no attempt is made to evaluate the scale of its influence or to address whether that influence is detrimental or not. Rather, if the project impacts are less than the local environmental standards then the effect is assumed to be acceptable.

**Notes**

1. A more in-depth discussion of EIA can be found in, e.g., Canter (1977), Munn (1979), O'Riordan and Hey (1976), O'Riordan and Sewell (1981).
2. The US National Environmental Policy Act of 1969, which requires the preparation of EIAs for 'major federal actions significantly affecting the quality of the human environment'.
3. Westman (1985), pp. 32–3 provides a (slightly outdated) list of the status of many countries' EIA systems.
4. This process is not peculiar to Japan. The European Community directive on EIA was also subject to constant modification and revision by various member states, which sought to weaken the proposed system.
5. Eversley, D.E.C. in O'Riordan and Hey (1976), pp. 130–1.

# Part one

# Environmental policy

# Chapter two

# Interest groups

## Participation in environmental policy-making

**Introduction**

Pollution in Japan has affected almost every aspect of the living environment and, with it, the health of the population. In December 1985, according to statistics produced by Japan's Environment Agency, almost 100,000 people were certified as suffering from pollution-related illness and entitled to compensation under the Pollution-Related Health Damage Compensation Law of 1969. The majority of these people suffer from respiratory illnesses linked to air pollution, such as chronic and asthmatic bronchitis.[1] Moreover in the past Japan has faced major pollution incidents, such as mercury poisoning at Minamata and Niigata and cadmium poisoning at Toyama, which caused death and permanent injury to thousands of people.

Surprisingly, despite this first-hand experience with widespread environmental problems, the Japanese have not turned 'green'. In fact, according to a United Nations survey of 1988, the Japanese are less worried about the environment and more unresponsive or hostile to environmentalists than people in almost any other country.[2]

In contrast, Japan is keen to adopt a leading role in the development of technology to deal with global pollution problems. In the 1970s Japan acquired a reputation for its development and use of such pollution-control devices as the catalytic converter and flue gas desulphurization units. Japan's renewed emphasis on technological measures for solving pollution problems is led by the Ministry of International Trade and Industry (MITI) and the ruling Liberal Democratic Party (LDP), which are undoubtedly attracted as much by the business opportunities this offers as by ecological motives.[3]

This chapter aims to examine the key interest groups involved in the environmental policy-making process in Japan and explain their level of environmental awareness. Understanding the attitudes of the players in the process is important, not only when examining the effectiveness of existing environmental policies (Chapter 5), but also in explaining current problems with EIA (Chapter 8). The following sections discuss individual interest groups and give two brief examples of how these interest groups interact on environmental issues.

### Interest groups

Figure 2.1 summarizes the interactions of Japan's main interest groups. The centre of the policy-making process is the national administration which is controlled by the LDP-dominated Diet (parliament). The national executive cooperates closely with Japan's industrial interests to ensure the continued development of the nation's economy. The central government's Environment Agency, local governments and the courts play a much weaker role and act as an interface between the general public and the national administration. Although citizen groups promote environmental issues, they are limited by general public apathy and by an unwillingness to address issues beyond local, health-related problems.

The interactions of these groups are conditioned by several factors. First there is a strong link between government and industry, both of which see economic growth as their primary goal. Second the Japanese government has traditionally been very stable: the LDP has dominated Japanese politics for the last three decades, and none of the opposition parties has even vaguely equivalent power. The relatively weak power

**Figure 2.1**   Interest group interactions

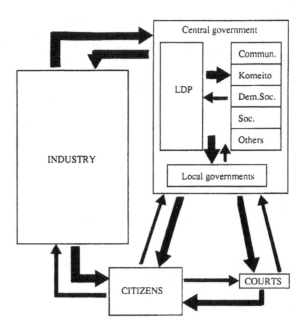

base of citizens, the Environment Agency's lack of influence, and the unwillingness of the courts to check the power of the administration and industry have reinforced the *status quo*. Thus Japan's social structure is extremely stable and its balance of powers is difficult to change.

However, recent trends indicate that change may be occurring. Major scandals involving bribery of high-ranking LDP politicians by industry, an alleged sex scandal, unpopular tax reforms and problems with agriculture policy led to the resignation of two prime ministers, Noburo Takeshita and Sosuke Uno, and seriously weakened the hitherto unquestionable authority of the LDP. Opposition to the LDP-industry complex is rising, and people are beginning to vote for the opposition parties.

Environmental awareness varies between the different groups and affects their approaches to environmental problems. The national government has historically tended to avoid serious consideration of environmental problems until they escalate to a national crisis.[4] Subsequent action has often been restricted by the administration's policy of promoting economic development. Although the government-industry link could theoretically lead to effective environmental protection, economic interests have generally prevailed to prevent environmental reform. Local authorities have been more sensitive to residents' opinions and have taken the lead in promoting environmental concerns. Industry and environmental groups have had a tragic history of conflict which climaxed with the 'Big Four' pollution lawsuits of the late 1960s (see 'Citizen groups' below and Chapter 3). An upsurge of public protest in the early 1970s, supported by the courts and the media, caused the government to institute far-reaching environmental reforms. The Environment Agency's role has recently been enhanced by the government's willingness to involve itself with global environmental problems. However, the courts have abandoned their pro-environment stance and once again support the government and industry despite pressure from the bar association.

Lately the public mood has moved away from environmental concerns and back towards economic revitalization. In part this is due to significant improvements in the levels of some pollutants which have caused many Japanese to feel that they have solved their pollution problems. It is also due to the pro-development mood set by the government–industry hegemony. Any environmental reform like EIA, which alters traditional policy directions and jurisdictional prerogatives, can be expected to be strongly opposed by the established interest groups unless some change in the existing power structure occurs internally or Japan is forced to change as a result of international pressure.

*Central government*

Legislative branch

Figure 2.2 outlines the organization of Japan's central government. Japan is a parliamentary democracy based on the British model. The chief legislative organ, the Diet, is composed of an upper and a lower house. The upper house or House of Councillors has 252 members who hold six-year terms of office. The lower house or House of Representatives has 512 members, who hold four-year terms. The prime minister is elected for two-year terms by Diet members. In turn, the prime minister chooses

**Figure 2.2**   Organization of the national government

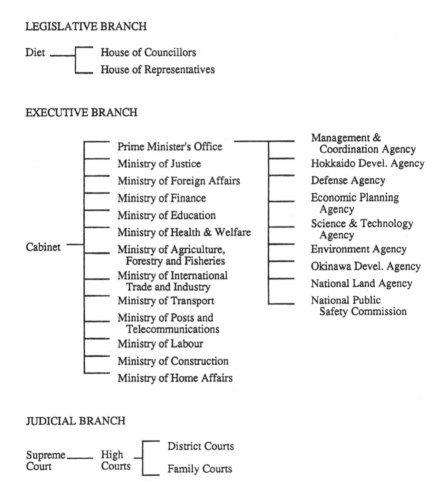

LEGISLATIVE BRANCH

Diet ─────┌── House of Councillors
          └── House of Representatives


EXECUTIVE BRANCH

Cabinet ─┬── Prime Minister's Office ───── Management & Coordination Agency
         ├── Ministry of Justice ── Hokkaido Devel. Agency
         ├── Ministry of Foreign Affairs ── Defense Agency
         ├── Ministry of Finance ── Economic Planning Agency
         ├── Ministry of Education ── Science & Technology Agency
         ├── Ministry of Health & Welfare ── Environment Agency
         ├── Ministry of Agriculture, Forestry and Fisheries ── Okinawa Devel. Agency
         ├── Ministry of International Trade and Industry ── National Land Agency
         ├── Ministry of Transport ── National Public Safety Commission
         ├── Ministry of Posts and Telecommunications
         ├── Ministry of Labour
         ├── Ministry of Construction
         └── Ministry of Home Affairs


JUDICIAL BRANCH

Supreme ──── High ──┌── District Courts
Court        Courts └── Family Courts

twenty Cabinet members, who also act as directors of the ministries.

Japan's political parties include the Liberal Democrats, Socialists, Democratic Socialists, Komeito, Communists and a few minor parties. The LDP has been in power since 1955. It held almost 60 per cent of the seats in both the upper and the lower houses (Figure 2.3) and virtually controlled Japanese politics until the upper house election of July 1989 gave the opposition parties a majority in the upper house for the first time (Figure 2.3). In the February 1990 lower house election, however, contrary to the expectation of many political analysts, the CDP suffered only minimal losses (11 seats) and retained overall control.[5]

The LDP has traditionally equated the good of society with economic growth, and as such its interests have been closely linked to those of industry. However, for the first time in its history this seems to have back-fired and become the root of the LDP's problems. In the 1988 'Recruit Scandal', high-ranking politicians were shown to have bought Recruit Cosmos shares prior to their sale on the open market, allowing them to make huge profits when reselling the shares later. This revelation led to a massive public outcry, the arrests of Recruit Co.'s owner, and the resignation of several senior ministers, including the PM.

The four major opposition parties normally appeal to only limited sectors of the population and seem unlikely independently to rise to

**Figure 2.3**  Diet party membership

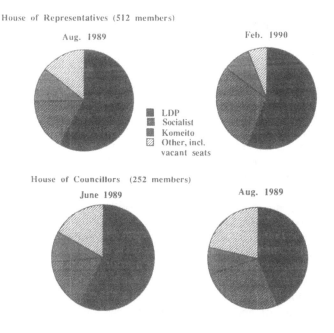

power. They are divided by deep ideological differences which makes coalition difficult.[6] Generally, however, they are less pro-business and more pro-environment than the conservative LDP. Recently the Socialist Party has profited from the public's anti-LDP sentiment, but the other opposition parties were less successful: the Democratic Socialists and Komeito were themselves tainted by the scandal, and the Communists suffered from a reaction against the massacre of students in Tiananmen Square in June 1989 and subsequent events in China.

Five environment-related parties (Anti-Nuclear, Greens, Earth Club, Environment Party, Life Party) were among the minor parties in the July election. Although restricted by a lack of candidates and resources, they nevertheless managed to gain 587,672 votes, roughly 1.3 per cent of all votes cast, and considerably more than in previous elections. This compares with the Greens' success in the European elections one month earlier, in which they obtained 15 per cent, 14 per cent, 11 per cent and 8 per cent of the vote in Britain, Belgium, France and West Germany respectively.

Towards the end of the 1980s it seemed that Japanese politics were at a juncture which could have led to the emergence of a two-party system. However, Japan's voters are notoriously conservative and Japan's institutions are extremely resistant to change. Although the Socialist party has recently made some gains, it has not successfully challenged the LDP's hegemony on Japanese politics. The LDP itself, although chastened by the Recruit Scandal, has not seriously set about the process of political reform nor changed its pro-development stance.

## Executive branch

The central government bureaucracy is composed of twelve ministries and a few minor bureaux and agencies which function as part of the prime minister's office. The LDP's long rule to date has precluded any major ministerial upheavals, and ministries are staffed by a self-perpetuating old boys' network from the most prestigious and established national universities. As a result, the ministries have become very powerful, very conservative and notoriously subject to factionalism and jurisdictional infighting. Within the bureaucracy issues such as environmental protection tend to be interpreted politically, in terms of the power shifts they cause among ministries, rather than in terms of their public benefit, and are thus fraught with political suspicions and recriminations. Environmental reforms particularly threaten the pro-industry MITI, Ministry of Transport (MoT), Ministry of Construction (MoC), and Ministry of Finance. These ministries are very powerful, and their opposition is the major barrier to such reforms.

Within the national administration, the Environment Agency (EA) has a relatively minor and poorly funded role as an intermediary between

**Figure 2.4**   Organization of the Environment Agency

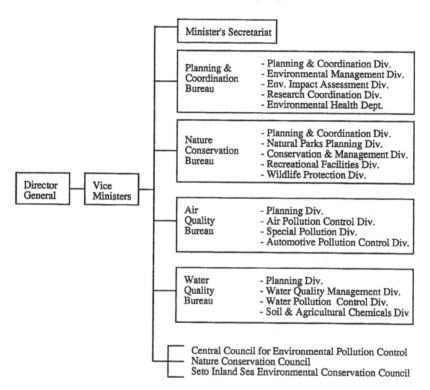

other government ministries and industry on one side and environmental groups on the other. The EA was established as part of the prime minister's office in 1971 in response to widespread public outcry against pollution. It was given the authority to deal with environmental issues previously scattered among a host of ministries, and the mandate to formulate and administer an integrated environmental policy. Figure 2.4 shows the EA's organization. The agency's Director General is appointed to the Cabinet by the PM, and has the same rank as a minister. The EA is divided into four bureaux: planning and coordination, nature conservation, air quality and water quality. It is advised by the Central Council on Environmental Pollution Control (CCEPC).[7]

Theoretically, the EA should be a centralizing force with enough political clout to get environmental considerations incorporated into national policy. In its early years it was, in fact, influential in enacting several environmental laws, and in setting and enforcing emission standards. However the EA's influence has been undermined by several factors:

— It is not the sole, or even the most powerful, agency responsible for the environment. MITI, the MoC, the Ministry of Health and Welfare (MHW), and the Ministry of Agriculture, Forestry and Fisheries (MAFF) all have jurisdiction over environment-related matters. None of these ministries have environmental protection as a first priority, and in some cases their primary objectives directly clash with those of environmental protection. This problem will be further addressed in Chapter 5.
— Enforcement of environmental policy is done by prefectural governors, although the EA has been known to exert pressure in cases of non-compliance.
— The EA has little influence over land use planning and control, which is the responsibility of the MoC and the National Land Agency. Municipal and prefectural governments are responsible for city plans and building permits, and various ministries are responsible for providing public facilities. Japan's land use planning system will be discussed further in Chapter 4.

The EA is also small and poorly funded. The EA's staff of about 900 is less than 5 per cent that of the MoT or MoC. Its 1987 budget of ¥47.3 billion is minute when compared with the MoC's budget of ¥4.6 trillion, or the ¥8 trillion allocated to transport-related projects. The MoT and MoC's environmental protection budgets alone are, respectively, almost twice and fifteen times the size of the total budget of the EA; however, as will be discussed in Chapter 5, these are spent primarily on sewage treatment systems and noise pollution control around airports.

The EA's low budget is only one manifestation of its low standing in the unofficial but crucially important hierarchy of ministries. This hierarchy is dominated by the Ministry of Finance, MoT, MoC and MITI. The EA's activities effectively limit the power of these ministries, which in turn are unwilling to give up their areas of jurisdiction to another, junior administrative body. The EA has no authority to present a bill to the Diet without the other ministries' consent, and is thus forced to seek their agreement and make compromises with them, further weakening its influence.

The EA is in the unenviable position of trying to cope with environmental problems in a highly political atmosphere. For instance it has been continually frustrated in its attempts to introduce legislation for EIA and the conservation of lakes and reservoirs.[8] As a result, it has retreated into the realms of technological rather than policy solutions, and essentially merely coordinates environmental policies rather than enforcing them;[9] the other ministries' close links to industry make them unlikely to perform the other functions adequately. Only the recognition that the world is facing a global environmental crisis, and increasing calls for Japan to take a major role in finding a solution to these problems will offer the opportunity for the future development of the EA. On 1 July 1990, a new 20-person World Environment Division was established in the EA; this may prove to be a brave new step in approaching these problems.

### Local government

There are forty-seven prefectures in Japan (see Figure I.1). Of these, Tokyo has special standing as a metropolis (*to*), Hokkaido as a 'circuit' (*do*), and Osaka and Kyoto as municipal prefectures (*fu*); the remaining forty-three are prefectures (*ken*), and all forty-seven together are referred to as 'to-do-fu-ken'.

The prefectures in turn are divided into municipalities: Japan contains about 600 villages (pop. <30,000), 2,000 towns (pop. <50,000) and 650 cities. In addition, ten 'designated cities' with more than one million inhabitants have been granted most of the power of prefectures (see Figure I.2). Prefectural governors and municipal mayors are elected in a manner similar to that of Diet members.

Although local governments are subordinate to the central government in other areas, they have been autonomous and active in the environmental field, and have generally preceded and set an example for the national government in terms of environmental policy. Local governments are in close contact with everyday environmental problems and receive residents' complaints directly. The actions of local governments are also not constrained by the influence of the pro-economic lobby to the same extent as central government actions. Between the late 1950s and early 1970s, in response to public pressure, many local authorities enacted their own pollution control regulations, some of which were stricter than national regulations. For instance, regulations produced by the Tokyo Metropolitan Government were later used as a guide when the central government tightened its own regulations. Local authorities are also leading the national government in the implementation of EIA procedures, as will be discussed in Chapter 6.

However, despite their leadership in many environmental issues, local governments are still strongly influenced by national policy. Local authorities get about 40 per cent of their revenue from the central government (plus 35 per cent from local taxes, and 25 per cent from bonds and miscellaneous revenues), so few of them can afford to directly oppose central government policy for fear of having these funds curtailed.[10]

*Industry*

Japan's economic sector is dichotomous. On one level are the politically powerful industries such as the steel, petrochemical, automobile and pulp/paper manufacturers. These giant conglomerates (*zaibatsu*) and central government have historically been closely linked: many *zaibatsu* were begun by the government in the late nineteenth and early twentieth century, and although they were later privatized, the government still has much control over their operation. The *zaibatsu* have powered the Japanese economy for the last century, and are afforded considerable status in their dealings with the administration. On the other level are the countless small firms and workshops which supplement and service the *zaibatsu*. These account for about two-thirds of all private sector employment and almost half of the GNP originating in the private sector.[11]

The industrial sector has a history of resistance to pollution control, even when pollution was causing severe human suffering. The *zaibatsu* are powerful enough to resist environmental reform, and the smaller firms have limited ability to invest in pollution control. This resistance is propounded by the government's lack of desire to enforce rigorous controls on industry. The government controls industry through informal 'administrative guidance' in the form of suggestions, requests, encouragement or warnings.[12] This method has the advantage of being flexible and often very effective. Administrative guidance also conforms with the traditional belief that consensus and cooperation is preferable to confrontation. However, such guidance is vulnerable to political influence and can be arbitrarily applied without recourse to review.

The effectiveness of administrative guidance also depends greatly on the cooperation of the parties involved. In the 1950s and 1960s administrative guidance was very effective when both industry and government shared the goal of revitalizing the economy. However when the government began to promote other aims, such as pollution control, industry reacted against guidance which it perceived to no longer be in its best interest. For instance the automobile industry, which cooperated closely with the government in the years after the war, has often opposed the government's emission standards, and even called for their abolition.[13] Lately, with the government's emphasis on economic revitalization, industrial and government aims are once again in closer alignment.

*Citizen groups*

The traditional Japanese reverence for authority and the tendency to subjugate personal wishes in favour of group harmony have been extensively discussed in other works.[14] Early citizens' groups tended to voice their protest in non-confrontational ways, through humbly phrased petitions or negotiation via a local elite. For instance, when vast areas of land at Ashio and Besshi were contaminated with residue from mining and smelting operations early in this century, the farmers petitioned the government, negotiated with the polluters, and in the end were treated fairly and compensated.[15]

However, traditional methods used for solving disputes in closely knit communities proved to be ineffective against the huge impersonal industries that sprang up after the Second World War. The tragic 'Big Four' pollution cases marked a turning point in methods of citizen protest. In the late 1950s the families of mercury poisoning victims in Minamata joined together to ask the Chisso Chemical Company to stop discharging mercury-laden wastes and compensate the victims. Five years later, residents of Niigata prefecture suffering from the same symptoms also formed a group to negotiate with the polluting company, Showa Denko Kanose. In Toyama prefecture, cadmium discharged into the Jinzu River by the Mitsui Mining and Smelting Company was diagnosed in the early 1960s as causing 'itai-itai' disease, and at the same time in Yokkaichi air pollution from several large *Konbinato* (industrial complexes) was causing or exacerbating respiratory problems.

In all these cases, the victims' polite petitions were met with the companies' denials of responsibility and with inconclusive government studies. The companies did pay a 'solatium'[16] to the victims, but this implied no responsibility and no apology, and imposed subtle social obligations on the victims to refrain from further demands.[17]

After years of desperate hardships the victims finally sued the companies in the late 1960s. The court decisions were decisive victories for the plaintiffs: they awarded substantial compensation and set a number of legal precedents for handling pollution-related disputes. Since the 'Big Four', lawsuits have become a much more acceptable method of resolving conflict in Japan, although they are still not often resorted to. Details of the 'Big Four' and other pollution-related injury-compensation suits are given in Table 2.1.

Over time, as popular understanding of environmental issues increased, and as Japanese people became more familiar with Western democratic institutions and confrontational methods, anti-pollution groups evolved from negotiation-oriented groups reacting to specific human health hazards to more litigation-oriented groups interested in preventing environmental harm. At present, citizens' environmental groups can be divided into three main types:

— Traditional pressure groups such as the Medical Association, the National Housewives' Association and labour federations have only a

**Table 2.1** Pollution-related diseases: victims and litigation

| Disease: location | Number of victims alive | dead | suing | Claim (in ¥ million) | Award (in ¥ million) | Date of filing | Date of verdict |
|---|---|---|---|---|---|---|---|
| Cadmium poisoning: |  |  |  |  |  |  |  |
| Toyama | 24 | 95 | 31 | 62 | 57 | Mar. 1968 | June 1971 |
|  |  |  | 34 | 151 | 148 | June 1971 | Aug. 1972 |
| Mercury poisoning: |  |  |  |  |  |  |  |
| Minamata | 1,385 | 752 | 138 | 1,500 | 937 | June 1969 | Mar. 1973 |
|  | (5,946) |  | 115 |  | 674 |  | Mar. 1987 |
|  | (25) |  |  |  |  |  |  |
| Niigata | 522 | 173 | 77 | 530 | 270 | June 1967 | Sep. 1971 |
| Respiratory diseases | 94,639 | 625 | 11 | 200 | 88 | Sep. 1967 | July 1972 |
| PCB in cooking oil: |  |  |  |  |  |  |  |
| Fukuoka | 1,578 | 51 | 44 |  | 680 | Feb. 1969 | Oct. 1977 |
| Kitakyushu |  |  | 729 |  | 6,080 | Nov. 1970 | Mar. 1978 |
|  |  |  | 342 |  | 2,500 | 1976 | Mar. 1982 |
| Quinoform/SMON: |  |  |  |  |  |  |  |
| Tokyo | ~ 11,000 |  | 35 |  | 870 | May 1971 | Oct. 1977 |
|  |  |  | 133 |  | 3,251 | May 1971 | July 1978 |
| Kanazawa |  |  | 16 |  | 431 | May 1973 | Mar. 1978 |
| Hiroshima |  |  | 43 |  | 1,070 | Apr. 1973 | Feb. 1979 |
| Chronic arsenic poisoning: |  |  |  |  |  |  |  |
| Toroku | 102 | 34 |  |  |  |  |  |
| Sasagadani | 8 | 13 |  |  |  |  |  |

Note: Parentheses signify the number of applications for official certification as pollution victims still pending in 1985. The numbers are taken from different sources and are continuously changing, and so should be taken only as an indication of the severity of the problem and size of compensation rather than as absolute figures. Other cases include poisoning by hexavalent chromium and thalidomide.
Source: Environment Agency (1987) *Quality of the Environment in Japan 1986*, pp. 186–94; McKean (1981) p. 47.

limited interest in environmental issues *per se* but represent a useful resource base for specific issues such as recycling or food contamination.

— Conservation groups are concerned with wildlife protection, preservation of scenic beauty and other traditional conservationist goals. Japan's largest conservation body, the Wild Bird Society of Japan, has 17,000 members, and the Nature Conservation Society of Japan has 7,000.[18]

— *Ad hoc* militant groups can be further divided into injury-compensation and anti-development groups. Some injury-compensation groups are listed in Table 2.1. Examples of anti-development groups are given in the case studies of Part 3.

The anti-development groups attract the most publicity and are those which most people associate with environmentalism in Japan. The Narita Airport opposition groups are particularly well known. When the MoT began building the New Tokyo International Airport (Narita) in 1966, both the farmers whose lands were appropriated and students who believed that Narita would be used for military purposes protested. The protests reached their peak in 1968, when five people were killed and 400 injured in riots. Construction was stopped for five years, and 1,500 guards are still needed to protect the airport from the missiles and bombs of radical protesters. Over the twenty-three-year dispute, almost 3,000 people have been arrested, almost all of them left-wing extremists and nearly 700 have been involved in trials related to the protest. In December 1989 the administration stepped up its activities against the protesters; for three days 6,500 riot police armed with water cannons stormed the protesters makeshift watch-towers and eventually toppled them. The MoT has now begun construction again, while at the same time clearing the last remaining resistance of the radical groups and wooing local farmers.

Narita is not the only example of such protest. Radical factions opposed to the construction of the Kansai International Airport were responsible for the bombing of several buildings and construction ships. These groups' reputation for violence and extremism has been detrimental to the image of environmentalism in Japan.[19]

Generally, however, environmental groups face apathy from the general public and from a government whose primary emphasis is on economic growth. Moreover, the environmental problems at stake now – nature conservation, improvement in quality of life, and international environmental problems such as acid rain and destruction of the ozone layer – are more difficult to solve, less visible, and have less personal impact than those of the late 1960s. Although Japan is not completely monistic the views held by those with influence tend to spread throughout the society. As such the public mood has moved away from environmental concerns towards wealth creation and consumerism.

The effectiveness of citizen groups in Japan is limited by two other factors. First, the movements cannot rightly be called environmental because they tend to focus on readily perceivable damage to human health and property rather than on wider environmental problems. Moreover, they link environmental quality mostly with limited pollution control measures and only infrequently with more fundamental social, political and economic questions.

Second, the groups are local, and make little effort to expand their circle of supporters or the scope of their campaigns. There are no strong national environmental pressure groups in Japan (although the anti-nuclear groups have begun an information network), so the local groups' actions generally remain uncoordinated and poorly funded. The groups initially avoided political affiliation, both to obtain a broader base of local support and because they believe that their existence resulted from the failure of the existing political system.[20]

Recently, however, citizen movements have become more active in the political arena. As mentioned earlier, environmental parties received about 1 per cent of the nation's votes in the last election. At the municipal level, the movements have had more success. In 1988 mayors were elected on anti-nuclear platforms at Kubokawa and Hikigawa. Four years earlier the mayor of Zusshi was elected because he opposed a proposed US military housing plan which could have a detrimental effect on the environment of the nearby Ikego Hills, and he was re-elected in 1988. More recently, in December 1989, the mayor at Rokkasho in Aomori Prefecture was elected on a platform to freeze the construction of the government's planned nuclear waste reprocessing plant.[21]

Environmental groups have also recently begun to create networks and share information. Japan's anti-nuclear movements have become the most successful of its present-day citizen movements, in part because several organizations provide liaison between the dozens of local movements, and publish a number of periodicals, booklets and books.[22] However these groups face the same internal limitations and external pressures as other movements.

## The courts

Disputes in Japan have traditionally been settled by negotiation, conciliation and consensus. Few cases are brought to court. Litigation is shunned not only because it is unfamiliar, expensive and time-consuming (see Table 2.1), but because there is strong social pressure to solve problems in a non-confrontational, non-public way. Furthermore, the Supreme Court decided in 1952 to rule only on specific issues, not on abstract constitutional concepts, and thus has limited its own role in reviewing administrative decisions.[23]

Nevertheless, the courts – primarily the lower courts – have played a major role with respect to environmental protection in Japan. During the 'Big Four' trials, they provided an unbiased forum for victims of pollution-related diseases, many of whom had been shunned and discriminated against before. Other court decisions since then have been supportive of environmental movements. More importantly, by gathering and disseminating complaints and by officially stating rules for resolving pollution-related conflicts, the courts made non-traditional methods for resolving conflict more acceptable.

To date, the courts have dealt primarily with two types of environment-related lawsuits: reactive cases concerning compensation for pollution-related diseases, and preventive litigation seeking to stop or modify development projects.

Reactive litigation is usually based on article 70 of the Civil Code, which provides for compensation in the event of injury or infringement of rights due to negligence.[24] The 'Big Four' lawsuits were of this type. In 1971 the Niigata district court ruled that Showa Denko Kanose had to pay compensation to residents who had contracted mercury poisoning from the wastes which it had dumped into the Agano River. Also in 1971, the Toyama district court concluded that the Mitsui Mining and Smelting Company was responsible for 'itai-itai' disease caused by its cadmium discharges. In 1972 the Yokkaichi branch of the Tsu district court found six firms guilty of polluting the air in Yokkaichi, and in 1973 the Chisso Chemical Company was required to compensate citizens of Minamata for the damages caused by mercury poisoning resulting from its activities. These lawsuits set important precedents for subsequent pollution-related injury-compensation suits:

— The burden of proof was shifted onto the industry. If it is likely that a toxic substance stems from a defendant's plant, the defendant must prove that this is not the case, rather than the plaintiff needing to prove that it is.
— New standards were set for proof of causation. Usually in tort cases a direct link must be shown to exist between the defendant's actions and the harm suffered by the plaintiff. However in cases involving certain pollutants and diseases, only a statistical, not a direct connection must be shown.
— Industries have joint and collective liability for multiple pollution sources, so plaintiffs can sue a single corporation rather than being obliged to make a case against all possible polluters.

The rulings also encouraged direct compensation negotiations between polluters and victims.

In recent years citizen groups have been less successful in their lawsuits. In November 1988, when the Chiba District Court ruled that the Kawasaki Steel Company should pay ¥77 million to 46 of 196 plaintiffs in an air pollution suit, the decision was heralded by environmentalists as a brake on the regression of environmental litigation and an important and heartening precedent for the pending decisions on similar lawsuits in Osaka, Nishi Yodogawa, Kawasaki and Kurashiki. Kawasaki Steel has appealed the decision.[25]

Preventive lawsuits for environmental protection have been of two types: civil injunctive suits and litigation concerning administrative action. The former seeks to prevent activities which may cause damage or injuries through pollution or destruction of the environment. The latter contests administrative decisions, usually under the Administrative Litigation Act of 1962; lawsuits based on EIA regulations would be of this type. Although several environmental groups have initiated lawsuits contesting administrative decisions, and although such suits are expected to play an important role in the future, they are still limited by a number of problems,[26] and injunctive suits are likely to continue being the primary legal method for environmental protection.

### The media and academic community

In the early 1970s television and newspapers were an important source of information on pollution incidents. In newspapers, the average space devoted to environmental problems grew from 0.4 per cent in 1960 to 2.8 per cent in 1972. News coverage quadrupled between 1965 and 1970.[27] The media at times blew environmental problems up to disturbing proportions, causing increasing public concern.[28] Lately, however, pollution issues no longer attract the same attention. Environmental protection measures are seen as limiting Japan's economic growth, and the attitude of the press has changed. Journalists are less sympathetic towards environmental groups and give more credence to the views of the politicians and the economic sector. Moreover, the majority of recent articles in the printed media concerned waste water and water pollution; long-term problems of nature conservation have been given less attention.[29] In 1989, however, global environmental issues seem to have taken up the dominant position in the attention of Japan's media with articles and features in most of the major magazines and newspapers.

Japan's academic community is very influential and yet has generally remained surprisingly uncritical of the handling of environmental issues. Academics are bound by a network of loyalties and obligations. Controversial fields of research are frowned upon, so they often remain rather conformist. Many academics are funded by and have links with industry or government research institutes. As a consequence, independence of thought and inquiry are limited. Nevertheless, some academics have proven to be leaders in environmental reform: for instance, Dr Noboru Hagino proved against great social pressure that 'itai-itai' is caused by cadmium poisoning, and Professor Ui Jun moved from the prestigious Tokyo University to Okinawa University rather than restrict his environmental activities.

### Examples of interest-group interactions

A sobering example of the interplay of Japan's interest groups is that of the pollution-related health damages compensation scheme. By 1986, compensation payments amounted to almost ¥100 billion, about five times more than had been anticipated. With air pollution levels falling, the Keidanren, Japan's biggest industrial association, lobbied for the system's abolishment. This move was contested by many organizations, including the Lawyers' National Information Exchange Committee on Pollution Cases, the Japan Scientists' Committee, the Japan Federation of Bar Associations and the Japan Environment Council.[30] However, under pressure from the industrial sector, in November 1983 the EA asked the CCEPC to study the future of the compensation system. It announced in October 1986 that air pollution no longer constituted the main cause of bronchial asthma and concluded that, on the basis of civil liability, there was no longer any reason to force polluters to pay compensation.[31] The designated pollution areas were cancelled and no new victims were certified after 1987, although presently certified victims will continue to receive payment.[32] This is a clear example of the relative weakness of the EA and its inability to defend its own policy instruments. Perhaps more worrying, it may also be the start of a far more dangerous trend – that of environmental deregulation.

Another example is the development of Japan's nuclear power programme. A public opinion survey of mid-1987 revealed that 86 per cent of 2,370 respondents were uneasy about nuclear power generation.[33] However a survey of Diet members undertaken a year later, the results of which are summarized in Figure 2.5, showed that most politicians' views did not reflect this concern. LDP members, who comprised about two-thirds of the Diet, generally believed that nuclear power is safe and should be expanded. With the exception of the Democrat Socialists, the opposition parties were less certain of its safety and did not favour expansion. If this trend can be extrapolated to other environmental issues, it suggests that a shift away from LDP rule could bring environmental policies more in line with popular beliefs.

**Figure 2.5**   Diet member views of nuclear power generation

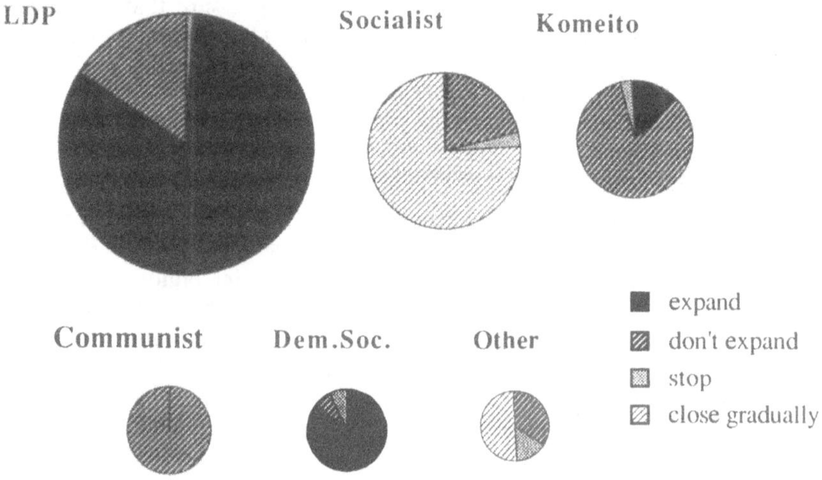

LDP          Socialist          Komeito

■ expand
▨ don't expand
▨ stop
▨ close gradually

Communist    Dem.Soc.    Other

Note: The area of the circles represents the parties' respective Diet memberships at the time of the survey.
Source: *Days Japan* (1 September 1988, in Japanese) 'Nuclear power: right or wrong', pp. 64–9.

### Conclusions

The current balance of interest groups, which has hindered the development of an integrated environmental policy in the past, could change rapidly in the next years. The establishment of regulations which allow citizens to challenge environmentally damaging developments (e.g. an EIA law) would strengthen the power of citizens, the courts and the EA, and weaken that of industry and some of the large ministries. Although the current hegemony of pro-industry ministries, the LDP and industrial interests makes the establishment of such regulations unlikely, recent changes in voting patterns have shown that this hegemony might weaken. If this happens, we are likely see the emergence of a broader environmental policy. The growing international awareness of environmental problems is also certain to influence the Japanese approach to environmental policy. The next five to ten years could represent a major transition period for Japan, with important ramifications for all developed and developing countries.

## Notes

1. Environment Agency (1988), *Quality of the Environment in Japan 1987*, pp. 188–9.
2. *The Economist* (15–21 July 1989), 'The first green summit', p. 119.
3. *Japan Times* (1 July 1989), p. 1.
4. Examples of this are incidents of severe soil pollution of farmland at Ashio and Besshi caused by mining operations in the early 1900s, and heavy metal poisoning at Minamata, Niigata and Toyama in the 1960s (see this chapter, 'Citizen groups' and Chapter 3).
5. International Cultural Association (1990), p. 10.
6. The Socialist Party, formed in 1945, is supported by labour unions. The Democratic Socialists developed in 1960 as a right-wing branch of the Socialists. The Komeito began in 1970 as the political wing of Soka Gakkai, a Buddhist organization. The Communists, formed in 1922, are strongly ideologically oriented with firm grass roots support and organization, and rely on funding from union dues and newspaper sales.
7. The CCEPC is composed of ninety part-time members divided into twelve subcommittees: general affairs, planning, pollution control programmes, cost allocation, environmental health, EIA, air quality, noise and vibration, traffic pollution, water quality, soil pollution, ground subsidence, and waste disposal.
8. Kihara (1981).
9. Gresser *et al.* (1981), p. 237.
10. Kiyoaki (1987), p. 99.
11. Yamamura and Yasuba (1987), p. 332.
12. Gresser *et al.* (1981), p. 233.
13. Tsukatani (1989).
14. E.g. Nakane (1986), Benedict (1986), Reischauer (1986), Kihara (1981).
15. Gresser *et al.* (1981), pp. 4–15.
16. The companies' offer of a solatium (*mimaikin*) instead of the compensation (*hoshokin*) demanded by the victims was significant. A solatium is a gift of sympathy for another's misfortunes; it does not imply responsibility. Compensation, instead, is paid as an open admission of responsibility, and is often accompanied by a public apology. The offer of a solatium was seen by the victims as the industry's attempt to calm protest while continuing to pollute, and was deeply resented: Huddle and Reich (1975), p. 120.
17. Ibid., pp. 102–32.
18. These compare with the British Royal Society for Nature Conservation and the Royal Society for the Protection of Birds, which have 165,000 and 500,000 members respectively, and the American Audubon Society with 550,000 members. Stewart-Smith (1987), p. 194.
19. *Japan Times* (13 Dec. 1989), p. 3. *Mainichi Express* (5 May 1985, 28 Jan. 1987, 28 Aug. 1987); Environment Agency (1988), p. 160.
20. Hase (1981a), pp. 34–5.
21. *Nuke Info Tokyo* (Mar./Apr. 1988), 'Kubokawa mayor abandons plan', 4, pp. 8–9; *Nuke Info Tokyo* (July/Aug. 1988), 'Hikigawa elects anti-nuke mayor', 6, p. 9; Van Brummel E. (spring 1986), 'Ikego hills: Zusshi citizens revolt against forest development for military housing', *Japan Environment Review*. *Japan Times* (12 Dec. 1989).
22. Hirose Takashi, a leading anti-nuclear activist, sees a new direction in the activities of environmental movements: '[People who] are moving now are not

just swept along. They are creating networks, trading information and ideas, strengthening each other. . . . People have given up on mass communication. . . . So it's not a movement in the old sense. It's a "work". . . . In fact nobody wants to build a "movement" . . . movements are always based on some martial metaphor. They're all out to "win", to gain a "victory". We're not aiming for a victory. . . . you don't kill off the opposition, you absorb it . . . you just inform, communicate': Kubiak, W.D. (Jan. 1989), *Japan Environment Monitor* 1(10), pp. 2–3.

23. Morishima (1981), pp. 82–3.
24. McKean (1981), p. 47.
25. *Japan Environment Monitor* (1988).
26. For instance:

    — What is defined as an administrative decision?
    — When is a case ripe for review? A Tokyo district court ruled against twelve plaintiffs who sought the repeal of a government licence to construct the Narita Shinkansen Line (Tokyo DC (23 Dec. 1972) 691 *Hanrei Jiho* 8). The court argued that when the licence was given the case was not ripe for review because no concrete sphere of interested persons could be defined under the proposed plan; the court made a distinction between the invasion of citizens' rights and a mere planning proposal. However, at the other extreme project plans can progress beyond the stage where they can be reviewed, and at that stage the project becomes virtually impossible to stop.
    — What is the definition of legal interest on the part of the plaintiff (as in the Narita Line case)?
    — What constitutes illegality (e.g. violation of established procedure, illegal exercise of discretional power, inaction): Kunamoto (1981). All of these problems also exist in other countries, as shown in Anderson *et al.* (1984).

27. Kelley *et al.* (1976).
28. Van Wolferen (1989).
29. Kelley *et al.* (1976), p. 140; Nomura *et al.* (1987).
30. Tsukatani (Sep. 1987).
31. Ibid., pp. 14–18.
32. Environment Agency (1988), p. 193.
33. Six per cent did not know, and only 8 per cent were not concerned: Prime Minister's Information Office (Aug. 1987), pp. 37, 54–5.

# Chapter three

# Environmental problems

## History and current state

### Introduction

Japan's environmental problems are integrally linked with its economic growth. Economic developments of the 1950s determined the nature of the severe pollution problems that faced Japan in the late 1960s and early 1970s. Similarly, present changes in Japan's economic structure will affect the future state of its environment.

In recent years, Japan has become a leader in pollution control technology and has acquired a reputation for improving its air and water quality at little harm to the economy. However, what remains are more entrenched environmental problems: afforestation, agriculture and industry have destroyed almost all of Japan's wild areas, with disastrous implications for wildlife and for the environment's ability to withstand stress and restore itself. Japan has a poor record on nature conservation, and despite improvements many areas are still badly polluted.

These problems are caused by a combination of political, economic and social factors. They cannot readily be solved, as was that of pollution, by the application of technology. How Japan deals with these problems will determine the quality of its environment as it enters the twenty-first century. The universal nature of the problems mean that Japan's solution will also be a model – or a warning – for other nations.

This chapter reviews the postwar history of environmental problems in Japan in relation to the country's economic growth. It presents four major paradigms:

(1) rapid postwar industrial development and economic growth (1955–63);
(2) the resulting severe pollution problems of the late 1960s leading to a rise in grass-root environmental consciousness (1964–9);
(3) the institution of a series of environmental regulations, and the OPEC oil shocks (1970–9); and
(4) decreasing pollution levels and an economic recession leading to increased emphasis on economic revitalization through massive public works projects (1980 onwards).

The chapter ends by postulating the repercussions which the last paradigm will have on future environmental and economic developments.

### Rapid economic growth (1955–63)

Japan has a high population and few natural resources. About 122 million people live on its $378,000km^2$, but since two-thirds of the country are uninhabitable due to steep mountains, the population density in its cities is among the highest in the world. Rice and other food crops are planted on every remaining flat patch of land, and much of the non-arable land is forested. About 70 per cent of Japan's land is forest, 20 per cent is agricultural, and almost all of the remainder is composed of cities and roads. Even so, Japan still imports about 30 per cent of its agricultural products and more than 60 per cent of its wood.

Lacking natural resources, Japan has specialized in importing raw materials, processing them, and selling the product. Table 3.1 summarizes Japan's foreign trade.[1]

The basis for Japan's present-day economy was established in the late 1940s and early 1950s. Having renounced the right to maintain military forces, Japan's national priorities centred around the development of energy resources and the re-establishment of the industrial base which had been badly damaged during the war. The first National Five-Year Economic Plan of 1955 proposed that an annual growth rate of 5 per cent be achieved by promoting industrial development, foreign trade and self-sufficiency. In 1960, prime minister Hayato Ikeda instituted the Ten-Year

**Table 3.1**   Exports and imports, 1985 (in $1 billion)

| Commodity group | Export | Import | Export/import |
|---|---|---|---|
| Food | 1.4 | 17.7 | 0.08 |
| Raw materials | 1.1 | 8.5 | 0.13 |
| Ore, minerals, metals | 1.6 | 11.4 | 0.14 |
| Fuels | 0.5 | 55.8 | 0.01 |
| Total primary products | 4.6 | 93.4 | 0.05 |
| Iron, steel | 13.6 | 1.5 | 9.1 |
| Chemicals | 9.1 | 8.4 | 1.1 |
| Other semi-manufactures | 4.8 | 2.9 | 1.7 |
| Office/telecom. equipment | 18.6 | 2.9 | 6.4 |
| Motor vehicles | 41.7 | 0.8 | 52.0 |
| Machinery | 45.3 | 8.4 | 5.4 |
| Textiles & clothing | 5.7 | 3.9 | 1.5 |
| Appliances & consumer goods | 30.6 | 3.6 | 8.5 |
| Total manufactures | 169.4 | 32.4 | 5.2 |

Source: General Agreement on Tariffs and Trade (1986), p. 149.

**Figure 3.1**   GNP growth (%)

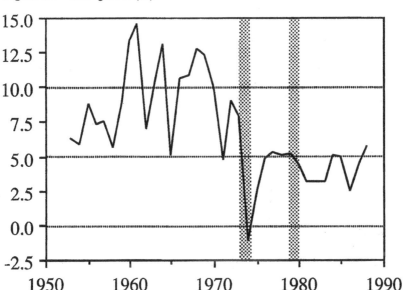

Note: Shaded areas indicate oil crises.

Income Doubling Plan, buttressed by new laws which designated twenty-one 'special industrial development regions' and which aimed to redistribute population and national income. These plans are discussed in more detail in Chapter 4.

A plentiful and inexpensive petroleum supply, an open international market, and close cooperation between the government and industry fostered Japan's economic boom in the 1950s and 1960s. The GNP rose by an average of about 10 per cent annually throughout the 1960s, as shown in Figure 3.1. The impact of the oil shock can be seen with GNP growth dropping to less than zero in 1974.

The remainder of this section summarizes the growth of some of Japan's major industries: energy, steel, automobile, ship, petrochemical/chemical and transport infrastructure.

*Energy*

One of the government's first priorities after the war was to establish a stable energy source for the manufacturing industries. Expansion of hydropower facilities was limited by the lack of suitable sites, and domestic coal was expensive and of poor quality. The ready availability of cheap Middle Eastern oil in the early 1950s caused a major restructuring of Japan's energy supply system: oil-fired power stations were rapidly built, and oil rather than coal became the primary energy source.[2] Electric power production, shown in Figure 3.2, doubled every seven years, rising from 45 billion kWh in 1950 to 476 billion kWh in 1975. Japan's total energy consumption also soared, tripling between 1955 and 1965, and doubling again by 1970.[3] By the mid-1970s, imported oil accounted for about three-quarters of Japan's primary energy; most of the rest came from hydropower and coal.[4]

*Steel*

Steel production, which had almost stopped during the war due to the shortage of coal, was one of the industries targeted in the government's postwar plans. With government assistance, steel producers underwent a series of modernization programmes from 1951 on, which aimed to bring Japanese steel prices down to international levels. Steel production boomed, as shown in Figure 3.3. Between 1950 and 1970, production doubled every five years, rising from 5 to 93 million tons, and peaking at 119 million tons in 1973. By 1960, Japan was the world's fifth largest steel producer, and by 1980 it was second only to the Soviet Union.[5]

*Automobiles*

Japan's automotive industry was given an early boost by American orders for army trucks for the Korean War. Commercial vehicle production grew steadily thereafter. Motorcycles and scooters were the cheaper and preferred form of personal transport until the late 1950s, but after that with rising incomes the domestic demand for automobiles grew rapidly: car ownership rose 100-fold between 1955 and 1975.[6] The overseas market for Japanese vehicles also expanded, from 0.2 million in 1965 to 1.1 million in 1970 and 6 million in 1980.[7] Automobile production, shown in Figure 3.4, more than doubled every two years between 1950 and 1970, and has been growing, albeit unevenly, since then.

*Ships*

Shipbuilding was also targeted as a priority industry after the war. With government assistance, the industry underwent a period of intense modernization in the early 1950s. The Korean War and the worldwide need for oil tankers provided an open market, and by 1956 Japan was the world's largest shipbuilder. Ever since, Japan has been producing about half of the world's tonnage of ships.[8]

**Figure 3.2**  Electricity production (billion kWh)

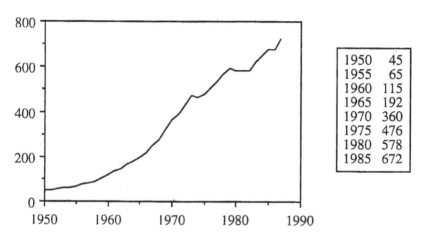

| 1950 | 45 |
| 1955 | 65 |
| 1960 | 115 |
| 1965 | 192 |
| 1970 | 360 |
| 1975 | 476 |
| 1980 | 578 |
| 1985 | 672 |

Sources: Mitchell (1982), p. 365; Agency of Natural Resources, MITI, various publications.

**Figure 3.3**  Steel production (million tons)

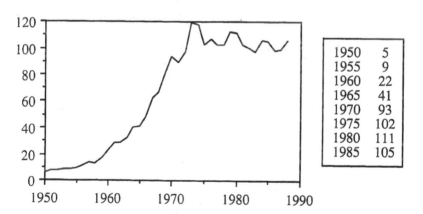

| 1950 | 5 |
| 1955 | 9 |
| 1960 | 22 |
| 1965 | 41 |
| 1970 | 93 |
| 1975 | 102 |
| 1980 | 111 |
| 1985 | 105 |

Sources: Mitchell (1982), p. 337; OPEC, various publications.

**Figure 3.4** Automobile production (1,000s)

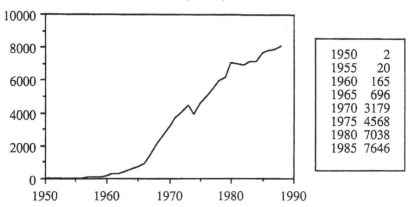

| | |
|---|---|
| 1950 | 2 |
| 1955 | 20 |
| 1960 | 165 |
| 1965 | 696 |
| 1970 | 3179 |
| 1975 | 4568 |
| 1980 | 7038 |
| 1985 | 7646 |

Sources: Mitchell (1982), p. 349; OPEC, various publications.

*Petrochemicals and chemicals*

The petrochemical and chemical industries also boomed. Petroleum refining for energy generation produced a surplus of naphtha, the raw material for the petrochemical production. In 1958–9, with government assistance, four companies built giant *kombinato* which combine oil refining, petrochemical processing, and sometimes power generation on one site to minimize costs. Five further *kombinato* were opened between 1960 and 1965. The production of ethylene, the main product of naphtha refining, rose from 78,800 tons/yr in 1959 to 733,000 tons/yr in 1965, and by 1977 fourteen naphtha centres and seventeen petrochemical complexes were producing 5.3 million tons/yr.[9] Production of other petrochemical and chemical products such as synthetic fibres, detergents, plastics, fertilizers and pesticides also rose as the technology improved and new markets opened.[10]

*Transport infrastructure*

The rapidly growing industries and increasing number of private cars required expanded transport facilities. In the late 1950s, the government embarked on several large projects to improve roads and airports.

Revenue from a gasoline tax was earmarked for highway projects in 1954, and a series of road improvement plans began in 1958. Between 1960 and 1975, paving on national highways rose from 29 per cent to 93 per cent. The first limited access expressway was opened in 1965, and by 1988 Japan had 4,400km of expressway.[11] As shown in Table 3.2, car

**Table 3.2**   Domestic transport

Passengers (in 1 billion passenger-km)

|  | Rail | Road | Ship | Plane | Total |
|---|---|---|---|---|---|
| 1950 | 105 (90) | 9 (8) | 3 (2) | — | 117 |
| 1960 | 184 (76) | 56 (23) | 3 (1) | 1 | 244 |
| 1970 | 289 (49) | 284 (48) | 5 (1) | 9 (2) | 587 |
| 1980 | 315 (40) | 432 (55) | 6 (1) | 30 (4) | 783 |
| 1985 | 330 (38) | 489 (57) | 6 (1) | 33 (4) | 858 |
| 2001 (est.) | 345 (30) | 732 (65) | 4 — | 53 (5) | 1,134 |

Freight (in 1 billion ton-km)

|  | Rail | Road | Ship | Plane | Total |
|---|---|---|---|---|---|
| 1950 | 34 (51) | 6 (9) | 26 (40) | — | 66 |
| 1960 | 55 (39) | 21 (15) | 64 (46) | — | 140 |
| 1970 | 63 (18) | 136 (39) | 151 (43) | — | 350 |
| 1980 | 41 (9) | 179 (41) | 222 (50) | — | 442 |
| 1985 | 22 (5) | 206 (47) | 206 (47) | 0.5 | 435 |
| 2001 (est.) | 12 (2) | 336 (65) | 166 (32) | 1.3 | 515 |

Notes: Numbers in parentheses denote percentages. Road traffic includes cars, buses, and lorries. The ratio of bus:automobile passenger transport fell from 8:1 in 1950 to 1:3 in 1980. Almost all freight is transported by lorry.

The ratio of private:national railway passenger transport rose from 1:2 in 1950 to 2:3 in 1980. Almost all freight is transported by national railway.

Some totals differ from the sum of transport modes due to roundoff error.

Source: Ministry of Transport, various publications.

usage quintupled between 1960 and 1970, and rose again by more than 50 per cent to 1985.

Airline passenger-km rose by more than 20 per cent per year between 1955 and 1975, and have continued to increase somewhat more slowly since then. A series of plans beginning in 1966 led to the expansion of air transport facilities, including construction of the Narita International Airport (opened 1978) and Kansai International Airport (to be opened 1993). These projects not only provided an infrastructure but also stimulated the economy through the purchase of concrete, steel and other domestic products.

## Recognition of pollution problems (1964–9)

This rapid industrial expansion led to an economic boom in Japan in the 1960s. The Ten-Year Income-Doubling Plan of 1960, which aimed for 9.1 per cent annual growth, actually achieved 10.7 per cent average growth, with a peak of 14.5 per cent in 1961. Rising private incomes led to increased spending, further stimulating the domestic market in an upward spiral of mass production and mass consumption.

The emphasis on economic growth ignored other problems. Japan's rapid industrialization changed its urban structure and traditional way of life; this will be discussed in Chapter 4. It also resulted in severe environmental pollution.

### Air pollution

Air pollution grew rapidly worse as a result of industrial emissions, home heating, and increased automobile use. Although the Tokyo Metropolitan Government had established a Soot and Smoke Prevention Law in 1955, and a similar law was enacted nationally in 1962, the laws applied only to limited regions with already high levels of pollution, were only minimally enforced, and proved to be ineffective in the face of the general emphasis on economic growth.

Total emissions of sulphur dioxide ($SO_2$, usually generated by the burning of fossil fuels) more than trebled in the 1960s, from less than 1.7 million tons/yr in 1960 to 5.8 million tons/yr in 1970. More than 90 per cent of this increase was attributed to greater fuel oil consumption. Nitrogen dioxide ($NO_2$, generated by the oxidation of atmospheric nitrogen during combustion, primarily in factories and automobiles) showed similar trends: emissions almost trebled during the 1960s, from 0.7 million tons/yr in 1960 to 2.0 million tons/yr in 1970. About 60 per cent of this increase came from greater liquid fuel consumption, and much of the rest came from vehicle emissions.[12]

It is difficult to reconstruct the precise extent of Japan's air pollution problems because monitoring was carried out only sporadically before the 1970s. Figure 3.5 shows changes in levels of indicative air pollutants. Annual average national $SO_2$ and carbon monoxide (CO) levels rose steadily until the late 1960s and have dropped since then. $NO_2$ levels have remained fairly constant with a slight increase in recent years. Incidences of excessive photochemical oxidants (generated by a reaction of hydrocarbons and nitrogen oxides initiated by sunlight) were numerous in the early 1970s, fell in the late 1970s, and have begun to rise again recently. The same is true of suspended particulate matter.

### Water pollution

Water pollution also became a serious problem as a result of growing industrialization and urbanization, insufficient waste-water treatment facilities, and increasing non-point source pollution from agriculture.

Industrial freshwater consumption rose from 10 billion tons/yr in 1962 to almost 16 billion in 1973.[13] Rapid postwar progress in such fields as

**Figure 3.5** Air quality

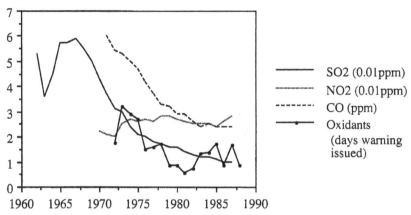

SO2 (0.01ppm)
NO2 (0.01ppm)
CO (ppm)
Oxidants
(days warning
issued)

Source: Environment Agency, various publications.

electronics and chemical engineering meant that the resulting effluents – which were often discharged into rivers untreated – not only grew in volume, but also contained unprecedented types of pollutants. In 1949, Tokyo enacted an Industrial Pollution Control Law in response to numerous complaints, and other prefectures[14] passed similar ordinances in the following years. In 1953, the Ministry of Health and Welfare drafted a pollution control bill, but opposition from government and financial interests prevented its enactment. Finally in 1958, after effluents from a Tokyo paper mill killed massive numbers of fish and irate fishermen stormed the mill, the national government passed the Industrial Effluent Control Law, Water Quality Conservation Law, and Sewerage Law. However these laws applied only to areas which were already heavily polluted, and their effectiveness was hampered by slow implementation and slack enforcement.[15]

Although sewage systems had been established in about fifty cities before 1940, most of these only transported waste-water without treating it. Human nightsoil was traditionally stored and used as fertilizer, and much remaining waste was discharged directly into the rivers. A Public Cleansing Law was enacted in 1954 and since 1963 a series of four-year sewerage construction programmes have been instituted. Nevertheless, in 1970 only 16 per cent of the population had been provided with sewage systems, and even in 1988 less than 40 per cent of the population were connected to sewer systems.[16]

Further pollution came from the leaching of agricultural chemicals into rivers and groundwater. The pesticides BHC and DDT[17] were introduced in 1945 and were soon widely used. Although both were banned in 1972, they take many years to decompose and are still a major source of the chlorinated hydrocarbons which pollute Japan's waters. Mercury compounds were used in insecticides and bactericides until they were banned in 1970. Chemical fertilizers were also used in vast quantities, and the nitrates and phosphorus compounds which leached from the fields led to the eutrophication of many water bodies, and to increasing numbers of red tide outbreaks. Japan is still one of the world's leading consumers of pesticides and fertilizers.

Whereas less than fifty water supply sources were contaminated in 1960, ten years later 583 were contaminated.[18] Figure 3.6 shows the percentage of water samples which comply with water quality standards relating to human health, and Figure 3.7 shows those which comply with standards relating to BOD and COD.[19] In 1970 water quality standards relating to human health were exceeded in 1.4 per cent of all samples, and in 1975 (the first year of monitoring) only 21 per cent of rivers, 38 per cent of lakes and 16 per cent of coastal waters met all the standards relating to the environment.[20] Coastal fisheries declined as a result of pollution, land reclamation and disruption from oil tankers. Treatment for drinking water became increasingly expensive, and in some cases heavy metal contamination caused water supplies to be cut off.

*Waste disposal*

The disposal of domestic and industrial wastes[21] became a pressing issue in the late 1960s as a result of dramatically increasing volumes of waste. With rising incomes, people generated more and different types of domestic wastes. Between 1966 and 1975, the volume of domestic wastes rose from less than 18 million to almost 32 million tons/yr; since then the volume has continued to increase, although less rapidly, to 38 million tons in 1985. Over time these wastes included an increasing proportion of plastics, which increases the difficulty of disposal. The cost per unit weight of disposing of domestic waste more than doubled every ten years, and the total cost of domestic waste disposal rose 100-fold between 1960 and 1980.[22]

Before 1960, kitchen refuse was collected separately from other domestic wastes and used as animal feed. More than thirty composting plants were built between 1956 and 1966, but at best they treated about 3 per cent of all wastes, and most were closed within fifteen years due to economic problems. Between 1963 and 1971 many incinerators were built, and the proportion of domestic waste which is incinerated has steadily risen from about 10 per cent in the 1950s to about two-thirds today. Almost all of the rest is disposed of in landfills. Many of the landfills are on reclamation sites, but this method of disposal is limited by the fact that garbage is not ideal fill material, and by the unsuitability of

**Figure 3.6**   Water quality relating to human health (% non-compliance)

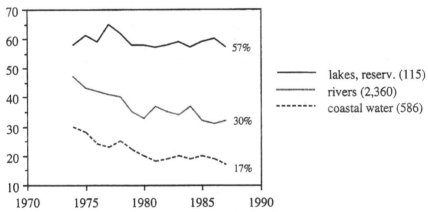

| 1971 | 0.63 | 1981 | 0.03 |
|------|------|------|------|
| 1972 | 0.28 | 1982 | 0.05 |
| 1973 | 0.23 | 1983 | 0.04 |
| 1974 | 0.20 | 1984 | 0.03 |
| 1975 | 0.17 | 1985 | 0.02 |
| 1976 | 0.09 | 1986 | 0.02 |
| 1977 | 0.08 | 1987 | 0.02 |
| 1978 | 0.07 |      |      |
| 1979 | 0.06 |      |      |
| 1980 | 0.05 |      |      |

Source: Environment Agency, various publications.

**Figure 3.7**   Water quality relating to environment (% non-compliance)

57%

lakes, reserv. (115)
rivers (2,360)
coastal water (586)

30%

17%

Source: Environment Agency, various publications.

many reclamation sites for storing wastes which may leach. Other sites are scarce. An experimental material recovery plant, Stardust '80, which separated pulp, metals and compostable materials for re-use and pyrolized the rest to produce fuel gas, operated for several years in Yokohama in the 1970s, but was closed down for economic reasons.

The volume of industrial wastes rose about eight-fold between 1955 and 1970, and stood at 292 million tons/yr in 1980.[23] Although about 60 per cent of this was recycled (including almost all metal scrap and wood chips), more than 100 million tons/yr of industrial wastes are still disposed of in about 10,000 facilities. About 90 per cent of these are facilities for waste treatment, 10 per cent are storage sites for non-hazardous wastes, and a small number are hazardous waste disposal sites.[24]

### Noise and vibrations

Noise and vibrations from traffic, airplanes, construction and factory operations accounted for more than a third of all pollution-related complaints in the late 1960s. Many of these problems are caused by the existence of factories and other businesses in close proximity to residential areas, which has been allowed by the comparatively weak system of land use planning. Mean noise levels in Tokyo in the late 1960s were 50dB in residential areas, 61dB in quasi-industrial areas, and 63dB in commercial and industrial areas, with much higher levels in locations close to airports and traffic intersections.[25]

### Ground subsidence

Subsidence (defined by the EA as a form of pollution) due to excessive groundwater pumping also continued despite the so-called Building Water Law of 1962. For instance in Tokyo water table levels dropped by about 10m between 1955 and 1965, accompanied by subsidence of $\geq$1cm/yr in more than half of the city, and of $\geq$10cm/yr in about 5 per cent of the city.[26] This continuous subsidence damages buildings, farmland and harbours, and increases the hazards of flooding.

### Pollution control efforts and the oil shocks (1970–9)

The widespread outcry against pollution, which began in the early 1960s and peaked with the 'Big Four' court cases in the early 1970s, prompted the government to establish one of the strictest and most effective anti-pollution programmes in the world. The hallmark of this programme was its use of the Polluter Pays Principle with no apparent reduction in GNP growth. The oil shocks of 1973/4 and 1978/9 also forced many industries to curtail energy use, a major cause of pollution, and make production methods more efficient.

Already in the early 1960s, citizens were protesting against the seemingly indiscriminate development taking place. When in 1963 Mishima-Numazu municipality became a 'special industrial development region' and local authorities approved plans for a new industrial complex, citizen protests were so strong that the local authorities had to withdraw their approval. In the national elections of 1964, environmental problems were the issue of greatest priority. Eisaku Sato, the prime ministerial candidate who promised to fight the pollution problems caused by unbalanced economic growth, won the election.[27]

The government gradually took action as pollution problems worsened and public criticism grew. MITI and the MHW set up environmental pollution control divisions in, respectively, 1963 and 1964. The Diet established a special committee for industrial pollution control in 1965, and in the same year the committee established the Pollution Control Services Corporation, which gave long-term low-interest loans to industries for the installation of pollution control equipment. Government budget allocations for pollution control increased rapidly after 1965.[28]

In 1967, the Diet enacted the Basic Law for Environmental Pollution Control as a master plan of future government environmental action. The Basic Law set out broad policy objectives and relative priorities, outlined methods for implementing this policy, and defined government and private responsibilities. It also provided for the establishment of environmental quality standards that would be 'in harmony with sound economic development'. However it proved to be ineffective due to the 'harmony clause' and a lack of enforcement provisions: pollution grew worse and, with it, citizen protest.

In 1970, a turning point in Japan's environmental policy was reached, prompted in part by a severe case of photochemical smog in Tokyo in July. PM Sato ordered his office's environmental pollution unit to unify the ministries' approaches to pollution control and to strengthen environmental legislation. In the extraordinary 64th Diet session of December 1970, the so-called 'pollution session', fourteen environment-related laws were enacted or amended, including:

— amendments to the Basic Law which removed the 'harmony clause';
— amendments which expanded the Water Pollution and Air Pollution Control Laws to include more types of pollution, establish nationwide standards rather than limited pollution control areas (eliminating the problem of pollution being controlled only in already severely polluted areas), and extend the pollution control powers of prefectural governors;
— enactment of the Pollution Control Works Cost Allocation Law, which requires developers to pay a portion of the cost of certain types of government-executed pollution control projects; and
— enactment of the Law for Punishment of Environmental Pollution Crimes Relating to Human Health, which penalizes individuals and corporations which discharge substances harmful to human health, and allows the use of presumptive evidence to prove cause-and-effect relationships.[29]

In February 1971, PM Sato's Cabinet proposed the establishment of an independent administrative agency to coordinate pollution control and environmental conservation programmes. That May, a law to establish the Environment Agency was passed through the Diet, and the EA was set up in July. In the following years, several important environmental regulations and standards were passed, including new ambient air quality and emission standards, the Nature Conservation Law (1972), the Pollution-Related Health Damage Compensation Law (1973), and the Chemical Substances Control Law (1973). Appendix B lists Japan's environmental legislation in chronological order.

Most importantly, the government showed a genuine willingness to counteract pollution, even at the risk of limiting economic growth. Provisions for balancing pollution control with economic growth were deleted from environmental regulations. No-fault liability was assigned to industries involved in pollution-related health damages, and the new emission standards, listed in Appendix A, were among the strictest in the world. The Polluter Pays Principle was applied very effectively through the various laws which charged polluters for their emissions, and for illnesses caused by their emissions. The current system of environmental policies and regulations will be further discussed in Chapter 5.

Another series of events had major implications for Japan's environment: the OPEC oil shocks of 1973/4 and 1978/9. In 1973, 73 per cent of Japan's primary energy was in the form of oil, 99.8 per cent of which was imported. Of the imports, about 85 per cent came from OPEC countries. This heavy dependence on imported oil caused Japan to be particularly hard hit when the OPEC countries limited exports and raised oil prices four-fold between 1973 and 1979. GNP growth dropped to less than zero in 1974, and has not since attained pre-1974 levels.

Government and industry response to the oil shocks can be summarized as follows:

— Oil and electricity consumption were reduced by 15 per cent in various industrial sectors, enforced by new laws[30] which allowed the government to regulate oil imports and consumption.
— Energy conservation became widely practised. The 1979 Law Concerning the Rationalization of Energy Use set up a programme of guidance and financial incentives for investing in energy-saving equipment and reducing energy consumption in factories, buildings, automobiles and electrical appliances. The government also sought to shrink energy-intensive industries which were no longer competitive. Between 1973 and 1987, the energy used per unit output fell by 27 per cent in the steel industry, 30 per cent in pulp/paper and cement manufacture, and 37 per cent in textile dyeing. Refrigerators' energy use fell by 67 per cent and car mileage increased from 9.6 to 11.8 km/l. In fact, industries established such stringent measures that further energy conservation is considered extremely difficult. As a result, industrial energy use has stayed steady since 1973 despite a 50

per cent rise in GNP, and *per capita* energy consumption in Japan is only 85 per cent of that in the UK, and 40 per cent of that in the US.[31]
— In response to the Petroleum-Substitute Energy Act of 1975, the energy industry shifted its emphasis towards nuclear power and other forms of power (solar, wind, wave) which can be produced domestically. This change is shown in Table 3.3. Between 1974 and 1979, about three nuclear reactors per year became operational, and since then about two reactors per year have come on line. Nuclear power production doubled between 1980 and 1987, and by mid-1988 thirty-six nuclear power stations were producing 28 per cent of Japan's electricity. By the year 2000, fifty-one nuclear reactors are expected to be operational.[32] As discussed in Chapter 2, these plans are very contentious.

It is difficult to differentiate between the economic effects of Japan's pollution control strategies and those of the oil shocks. Certainly GNP growth dropped, and certain industrial sectors suffered. However, the consensus seems to be that anti-pollution measures have, if anything, helped Japan's economic growth. While the cost of certain goods rose to reflect the cost of installing pollution control equipment – e.g. pulp/paper by 5.6 per cent, chemicals by 3.7 per cent, metals by 3.3 per cent – and thus the demand for those goods declined, the demand for pollution control equipment and peripheral services increased. Overall the positive economic effects of pollution control seem to have exceeded the negative effects.[33]

The combination of strict pollution-control measures and reduced energy consumption rapidly led to lower pollution levels. Annual average $SO_2$ levels fell by about 80 per cent, CO levels by about 60 per cent, and the percentage of water samples which did not meet human health

**Table 3.3** Energy supply

|  | 1977 | 1987 | 2000 (est.) |
|---|---|---|---|
| Oil (mil. kl) | 311 (75) | 260 (57) | 242 (45) |
| Coal (mil. tonnes) | 78 (15) | 108 (18) | 136 (19) |
| Nuclear (GW) | 8 (2) | 28 (10) | 54 (16) |
| Nat. gas (mil. kl) | 8 (3) | 44 (10) | 58 (11) |
| Hydro. (GW) | 26 (5) | 38 (4) | 46 (4) |
| Solar, wind, etc. (mil. kl) | 0.3 | 6 (1) | 25 (4) |
| Geothermal (mil. kl) | 0.1 | 0.4 | 4 (1) |
| Total (in mil. kl) | 412 | 457 | 540 |

Note: numbers in parentheses denote percentages.
Source: MITI Agency of Natural Resources and Energy, various reports.

standards fell by 98 per cent. Levels of other air pollutants stayed about even despite rising GNP, as did the percentage of water samples which did not meet environmental protection standards. These improvements were widely publicized, and Japan gained a reputation for having the world's most stringent emission standards and most advanced pollution control equipment.

However, other problems remained. Nature conservation did not gain popularity due to insufficient government funding and the belief that conservation is not appropriate for Japan.[34] Water and soil pollution, more difficult to combat technologically than air pollution, remained. Many of the more polluting industries set up factories overseas, and nuclear wastes were sent to the UK for processing. Quality of life in the cities remained poor. Most importantly, however, concern about environmental issues fell as concern about the economy rose.

### Economic recession and the current state of the environment (1980–90)

The oil shocks led to a decline in the manufacturing sector, a rise in the service and information sector, and corresponding changes in the employment structure. Although pollution which can harm human health was radically reduced, improvements in the quality of life and the natural environment lag behind, since such improvements require a change in consumer habits and living patterns rather than technological improvements.

Since the mid-1970s the dominant postwar industries have generally not grown at the same rate as before, and some have drastically curtailed production. The reasons are multiple: increased oil prices, the worldwide recession, competition from rapidly developing countries like Taiwan and South Korea which (still) have lower labour costs, and increased protectionism from the US and European countries which feel that Japan has been too eager to grab market shares without opening up its own market.

The steel and the shipmaking industries have been hit by world over-supply from cheaper producers. As shown in Figure 3.3, steel production dropped from a peak 119 million tons in 1973 to about 106 million tons in 1988,[35] and even this level is dependent on the government's recent expenditure on public works projects. The future of shipbuilding looks even grimmer: only four years after the 1975 peak production of 17 million tons, production was down to 4.7 million tons, and in 1987 it stood at only 8.2 million tons.[36]

Automobile production is stll growing, but more slowly than before. More than 8 million cars were produced in 1988, compared with slightly over 7 million in 1980 and 4.6 million in 1975 (Figure 3.4), and Japan is still the world's biggest automobile manufacturer. However, exports have

**Table 3.4**  Changes in workforce ratios (%)

| Sector | Year | | | | | | | |
|--------|------|------|------|------|------|------|------|------|
|        | 1950 | 1955 | 1960 | 1965 | 1970 | 1975 | 1980 | 1985 |
| Primary | 48 | 41 | 33 | 25 | 19 | 14 | 11 | 9 |
| Secondary | 22 | 24 | 29 | 32 | 34 | 34 | 34 | 33 |
| Tertiary | 30 | 36 | 38 | 43 | 47 | 52 | 55 | 58 |

Note: The primary sector includes agriculture, forestry, and fishing; the secondary sector includes mining, construction and manufacturing; the tertiary sector includes sales, finance, transport/communication, services and government.
Source: Bureau of Statistics, Management and Coordination Agency, 'National Census'.

dwindled due to an appreciated yen and increasing market protectionism. In response, Japan has increased its overseas production to combat high domestic labour costs.

The production of chemicals and petrochemicals has also lagged behind the optimistic forecasts of the early 1970s, due primarily to increased oil prices. Naphtha production has declined steadily since the mid-1970s. Ethylene production, which stood at 5.3 million tons in 1977 and was predicted to be 15 million tons in 1985, was only 4.3 million tons in 1986. The production of other chemicals also dropped sharply after the oil shocks but have otherwise risen slowly.[37]

However, whereas the secondary (manufacturing) sector seems to be stagnating after its rapid postwar expansion, the tertiary (information and service) sector is expanding rapidly, as shown in Table 3.4. The information industry, which accounted for 7 per cent of the GNP in 1988, is expected to grow to about 20 per cent by the year 2000.

Biotechnology and new materials are also expected to be growth areas. Biotech developments are envisaged in the chemical, medicine, food, fibre, paper/pulp and research industries, and the market for biotech products is expected to rise more than twenty-fold between the mid-1980s and the year 2000.[38]

At present, Japan's air pollution levels are kept low by the strict pollution control measures of the 1970s, and the lower fuel consumption and depressed smokestack industries of the 1980s. As shown in Figure 3.5, $SO_2$ levels have dropped from a 1967 peak of 0.059ppm to about 0.012ppm in the mid-1980s. Compliance with annual and daily $SO_2$ standards was, respectively, $\geq$ 99 per cent and $\geq$ 98 per cent in the mid-1980s. CO levels dropped from 6ppm in 1970 to about 2.4ppm in the mid-1980s, meeting all standards.

However, $NOx$ levels have remained steady at about 0.024ppm, replacing $SO_2$ as the main pollutant: $NO_2$ standards were exceeded at about 3 per cent of air pollution monitoring stations and 27 per cent of automobile exhaust monitoring stations in the mid-1980s. In addition, photochemical oxidant warnings rose in the mid-1980s after falling in the early 1980s. Concentrations of suspended particulates have fallen, but only about 50 per cent of monitoring stations meet particulate standards.[39]

Since the 1970s, water quality has greatly improved in terms of human health standards, and only 0.03 per cent of all samples currently do not meet the standards. However, water quality with respect to the living environment is still a serious problem. In the mid-1980s, about one-third of the water areas sampled failed to meet environmental water quality standards for BOD or COD, including almost 60 per cent of lakes and reservoirs. Closed water areas are progressively eutrophying, and water quality is falling due to the build-up of nitrogen, phosphorus, and algae. Many rivers, especially those around Tokyo, have BOD levels near or above 5ppm, a level above which most fish have trouble surviving. The coastal areas are threatened by red tides caused by eutrophication.[40]

Noise pollution levels have shown no improvement. On average 22,000 complaints are received each year, mainly against factories and other businesses (61–7 per cent) and construction work (~13 per cent). Complaints against motor vehicles and aircraft are only 2–4 per cent each, presumably because complaints against non-specific offenders are unlikely to be effective. In 1984 only 15 per cent of all automobile noise monitoring points (and only 4 per cent of those in residential areas) met noise standards. Although airplane noise is controlled by quieter engines, specified flight routes and times, and noise-proofing on houses, residents around airports are increasingly bringing lawsuits or appeals for compensation against airports.[41]

Ground subsidence has slowed down in the large cities after restrictions on groundwater pumping were imposed, but about 700km$^2$ are still subsiding by $\geq$ 2cm/yr throughout Japan. In 1985, soil pollution by heavy metals requiring countermeasures affected 126 farms covering 6,900ha. Soil pollution in urban areas recently gained attention when toxic substances were detected at sites formerly used for industrial purposes.

Finally, Japan's record on nature conservation remains poor. Although about 5.3 million ha (14 per cent of Japan's land area) have been designated as national and prefectural parks, only 5,600ha are wilderness areas and less than 30,000ha are wildlife protection areas.[42] Japan remains the world's largest trader in endangered species.[43] Agricultural productivity is maintained through single-crop plantings and extensive applications of herbicides and pesticides, but these destroy natural predator–prey relationships and the species variety needed for ecosystems to restore themselves.

The current state of Japan's environment reflects a number of limitations in the system of environmental protection. First, the system is heavily oriented towards preventing harm to human health, but puts much less emphasis on protecting the natural environment. This can be seen by Japan's relative success in meeting the different pollution standards relating to human health and the living environment. As will be shown in Chapter 5, the budgets allocated to these two goals also reflect their respective importance.

Second, Japan's emphasis has been on technological solutions rather than on solutions which require a change in social attitudes. For instance $SO_2$ emissions have been very effectively controlled through the installation of flue gas desulphurization units, but a decrease in NOx emissions, which would require decreased use of automobiles (a status symbol), has not been possible.

Third, Japan's system of environmental protection emphasizes after-the-fact measures for solving and redressing pollution problems (e.g. compensation for pollution victims, installation of pollution control devices) rather than pollution prevention. Japan's land use planning system only minimally considers environmental factors, and environmental impact assessment is not mandatory. Such an approach does not solve environmental problems but merely treats their symptoms. It has the additional drawback of causing a long-term decrease in the quality of life, as discussed in Chapter 4.

Solving these problems requires changes in social values and attitudes. Such changes could include:

— the development of manufacturing processes which produce fewer pollutants;
— reduced use of disposable containers and more recycling;
— reduced car use, encouraged by improved public transportation, the construction of cycle paths and pedestrian centres, and a freeze on new highway construction; and
— increased consumption of organic produce.[44]

All of these require individuals to consciously change their lifestyles, whereas at present consumption and ownership of luxury goods are encouraged.[45] As discussed in Chapter 2, the hegemony of interest groups which favour economic development, the powerlessness of those which favour environmental protection, and the general public's lack of concern about broader environmental issues make it unlikely that environmental policies will be integrated into the administrative planning process. This is particularly obvious when one considers the plans currently being put forward for Japan's high-tech future.

### Japan's high-tech future vision

Rising concerns about the state of Japan's economy and decreasing fears about its environment caused priorities to shift towards economic revitalization in the mid-1980s. Japan's development plans have consistently advocated the redistribution of industry and population away from the large metropolitan areas and towards less economically buoyant regions, the development of infrastructure and the stimulation of the domestic economy.

These goals have led to the creation of a high-tech future vision for Japan which has been promoted by developers and enthusiastically adopted by the government. The scheme involves a network of airports, high-speed expressways, and Shinkansen ('bullet train') lines which would make all parts of Japan accessible within 90 minutes. As a first step, this vision involves the construction of several major public works projects, including:

— the Kansai International Airport (to be completed in 1993);
— a highway across Tokyo Bay (to be completed by the late 1990s) and several artificial islands in the bay;
— three series of bridges linking Honshu and Shikoku islands (1965–99); and
— expansion of the expressway network to 14,000km by early in the twenty-first century.

These projects will be discussed at greater length in subsequent chapters.

The thinking behind the projects, however, deserves a brief comment. There is little evidence to suggest that these developments are the best option for boosting Japan's economy, especially in the long term. Major public works do little to improve the average person's quality of life, and their construction has a severe impact on the environment. In addition to irreparably damaging the immediate site on which they are located, the proliferation of airports and highways also encourages further air and noise pollution. The environmental implications of this proliferation of large-scale projects have also not been adequately examined: Japan's primarily reactive system of environmental protection is unable to prevent or even adequately mitigate the projects' environmental impacts.

An alternative approach would be to consider ways to develop the existing infrastructure while encouraging modest growth and environmental preservation. For instance, expansion of public transport, sewage systems and waste treatment facilities would produce jobs, strengthen the infrastructure and prevent further environmental degradation. However these ideas are not in vogue in Japan. An EA official recently noted that 'nowadays if a politician wants to promote environmental conservation, he will lose votes because people want economic activity'. The concept that environmental conservation harms economic activity is very dangerous, especially considering the delicate state of Japan's environment.

## Conclusions

After the war, Japan rose rapidly to economic power, helped by low petroleum prices, an open international market, and government–industry cooperation. Severe environmental problems of the 1960s caused widespread public outcry and forced the government to adopt some of the strictest pollution control measures in the world. These measures, and a worldwide recession brought about by the OPEC oil shocks, resulted in dramatic reductions in the levels of some pollutants in Japan. Public concern consequently moved away from the environment and toward economic revitalization.

Ideally, one would stop here, reconsider carefully the events of the last three decades, and devise a strategy which permits both environmental protection and economic growth. However, this has not been done: the government/industry hegemony equates the national good with rapid economic growth, people have been unwilling or unable to change their values, and environmental concerns have been limited to technological measures to reduce the levels of pollutants which affect human health. Japan is heading towards more large-scale development without assuring that such development is in the country's long-term environmental or social interest.

Although major pollution incidents like those of the 1960s are unlikely to reoccur, environmental conditions in Japan are likely to deteriorate unless major social and political changes occur to shift priorities away from short-term economic growth towards a more stable balance of economic and environmental concerns.

## Notes

1. By 1988, both imports and exports had dropped by about 25 per cent, but the ratios of imports to exports, and of primary to manufactured products remained generally stable.
2. Uchino (1978), pp. 70, 92.
3. Total energy consumption has increased as follows (in 1 trillion Kcal): 1955 – 513; 1960 – 844; 1965 – 1,458; 1970 – 2,841; 1975 – 3,411; 1980 – 3,730. Cited in the MITI annual report of industrial statistics.
4. *Kodansha Encyclopedia of Japan* (1985), vol. 2, p. 214.
5. Mitchell (1982) p. 337; *Kodansha Encyclopedia of Japan* (1985), vol. 3, pp. 334–5.
6. Uchino (1978), p. 162; *Kodansha Encyclopedia of Japan* (1985), vol. 7, p. 102; Mitchell (1982) p. 349. Further automobile ownership in Japan is limited by urban congestion, lack of parking facilities and the high cost of petrol.
7. Japan Transport Economics Research Centre (1982), p. 37.
8. Ibid., p. 38; *Kodansha Encyclopedia of Japan* (1985), vol. 7, p. 142.
9. *Kodansha Encyclopedia of Japan* (1985), vol. 6, p. 180.

10. Bank of Japan (various years), 'Comparative international statistics'.
11. *Kodansha Encyclopedia of Japan* (1985), vol. 7, p. 102; *Japan* 457 (Nov. 1988), p. 3. The tenth road improvement programme of 1988 calls for extension of the expressway network to 6,000km by 1993.
12. Uchino (1978), p. 166.
13. This does not include recycled water. *Kodansha Encyclopedia of Japan* (1985), vol. 2, p. 227.
14. Kanagawa, Osaka and Fukuoka prefectures.
15. Hashimoto, M. (1985), p. 123.
16. Environment Agency (1987), *Quality of the Environment in Japan 1986*, p. 123. This compares with 97 per cent and 72 per cent in England and the US.
17. Benzene hexachloride and dichloro-diphenyl-trichloro-ethane.
18. Kelley *et al.* (1976), p. 85.
19. Biochemical oxygen demand (BOD) and chemical oxygen demand (COD) are commonly used indicators of water quality. Lower BOD/COD levels indicate higher quality.
20. *Kodansha Encyclopedia of Japan* (1985), vol. 2, p. 227. Standards for the living environment include standards for BOD, COD, bacteria concentrations, suspended solids, etc. If only BOD and COD are considered, 57 per cent of rivers, 39 per cent of lakes and 72 per cent of coastal waters met the standards in 1975.
21. The Waste Disposal and Public Cleansing Law of 1971 (which replaced the Public Cleaning Law of 1954) distinguishes between general and industrial wastes: municipalities are responsible for collecting and disposing of general wastes (domestic and some industrial), but industrial wastes are to be disposed of by the dischargers.
22. Hiraoka (1986), p. 11.
23. Of this, 30 per cent was sludge, 21 per cent slag, 17 per cent livestock excretions and 10 per cent demolition wastes. Environment Agency (1987), p. 170.
24. Environment Agency (various years), *Quality of the Environment in Japan*, Tokyo; Trade and Industry Technological Data Research Co. (1975), pp. 219–24; Hashimoto, M. (1985), p. 31.
25. Tokyo Metropolitan Government (1971), pp. 150–4. For comparison, a very quiet room is about 35dB, and an automobile at 7m is about 70dB. EQS for noise set in 1971 are, e.g.: for residential areas, 50dB (daytime), 45dB (morning/evening), 40dB (night); for quasi-industrial areas 5dB higher for each category; for areas facing roads with two lanes 5dB higher for each category.
26. Ibid., pp. 197–214.
27. Hashimoto, M. (1985), p. 15.
28. Ibid., pp. 12–17.
29. Ministry of Foreign Affairs (1973/4), pp. 5–10.
30. The Oil Supply and Demand Adjustment Act, and the Emergency Act for Stabilizing the National Livelihood, both enacted in December 1974.
31. Ministry of International Trade and Industry (Dec. 1988), pp. 12–13; Weidner (1986), p. 53; Foreign Press Centre (1987), pp. 61–2.
32. Ministry of International Trade and Industry (Dec. 1988); Japan Electricity Production Information Centre (1988); Martineau, L. (12 Jan. 1987), 'Japan to rely on N-power', *Guardian*.
33. Corwin (1980), pp. 154–7; OECD (1977), pp. 86–7; Weidner (1986), p. 337.

34. Stewart-Smith (1987), p. 201.
35. Mitchell (1982), p. 337.
36. In 1975, 946 ships totalling 18 million tons were produced. Japan Transport Economics Research Centre (1982), p. 38; *Kodansha Encyclopedia of Japan* (1985), vol. 7, p. 102; OECD (monthly reports), *Main Economic Indicators*, Paris.
37. Japan External Trade Organization (1987, 1988), *Business Facts and Figures*; Kelley *et al.* (1976), p. 37.
38. Ibid.
39. Environment Agency (1987), pp. 77–106.
40. The number of red tide outbreaks has remained even at about 200/yr since 1979, after rising from 60/yr in the mid-1960s to more than 300/yr in 1976; ibid., pp. 67–9, 111–18.
41. Ibid., pp. 139–66.
42. Ibid., pp. 196–9.
43. Stewart-Smith (1987), p. 179.
44. These suggestions do not apply only to Japan, but to most countries.
45. For instance, in 1988 Kyoto City began a campaign against bicycle parking in central city areas, because it was felt that bicycles degrade the character of the neighbourhood; car parking was not banned.

# Chapter four

# Environmental management
## Economic and land use framework

### Introduction

Economic development and land use plans affect the type and extent of development which takes place in a given area, the spatial distribution of population and activities in that area, and thus the area's environmental quality. The establishment of an environmental management system, of which EIA would be one component, is restricted in Japan by conflicts between existing management/planning systems and possible reform programmes.

This chapter reviews land use and planning problems in Japan. It outlines the institutions responsible for, and the framework of, existing economic and land use planning systems. It argues that these systems do not fulfil their objectives of easing urban congestion, regional disparities and the negative side-effects of development. The role of environmental management systems is also discussed. Japan's planning system is highly developed, complicated and difficult to summarize; this chapter emphasizes main policy directions and mechanisms for their achievement.

### Land use and planning problems

Since the 1950s, much of Japan's rural population has moved to the cities. In 1950, only 37 per cent of Japanese people were living in areas officially designated as cities (*shi*); this proportion increased to 63 per cent in 1960, 72 per cent in 1970, and 76 per cent in 1980. The proportion of the population living in Densely Inhabited Districts, which correspond to actual built-up urban areas rather than to administrative divisions, rose from 44 per cent in 1960 to 54 per cent in 1970 and 60 per cent in 1980.[1] As shown in Table 4.1, urban areas and roads have steadily grown while agricultural, forest and moorland areas have shrunk. Between 1950 and 1960, the rural population dropped by more than one-third, while the urban population almost doubled.[2] Japan effectively changed from a predominantly rural into a predominantly urban society in just a few decades.

**Table 4.1** Land use changes (%)

|             | 1965 | 1975 | 1985 |
|-------------|------|------|------|
| Urban       | 2.5  | 3.6  | 4.4  |
| Roads       | 2.4  | 2.9  | 3.3  |
| Agriculture | 19.0 | 17.1 | 18.1 |
| Moorland    | 1.9  | 1.2  | 0.8  |
| Forest      | 74.2 | 75.2 | 73.4 |

Source: Siman (Mar. 1989), p. 14.

In turn the cities experienced a population boom, as shown in Table 4.2. Between 1960 and 1980, the population of the Tokyo Metropolitan Region rose by more than 60 per cent, and its share of the total population rose from 17 per cent to 23 per cent. During the same time the populations of the Osaka and Nagoya metropolitan areas rose by about half. Japan's population continues to grow by about 700,000 people annually, although this number is slowly falling.

Infrastructure provision in the cities could not keep up with the rapid rise in demand caused by the population boom. As a result, Japanese cities lag behind those of other developed countries in the provision of sewerage, transport facilities, green areas, sport and cultural amenities, and other infrastructure; this is shown in Table 4.3.

Land prices in the cities have also boomed as a result of the increased demand for office space.[3] The government has attempted to cope with this demand by easing housing standards to promote the construction of private houses, but this has caused residential land prices to soar. Between 1955 and 1984, residential land values in Japan increased by a factor of more than fifty; consumer prices, in contrast, rose by a factor of five.[4] Sales of public land in the cities have also increased the price of nearby properties. Private homes in the cities are sold for enormous sums, and the profits are used to buy more land and larger houses in the suburbs.

However, as people moved away from the cities to avoid high prices,

**Table 4.2** Population of greater metropolitan areas (in millions)

|          | 1960 | 1970  | 1980  | 2000 (est.) |
|----------|------|-------|-------|-------------|
| Tokyo    | 16.5 | 22.8  | 27.3  | 30–2        |
| Osaka    | 9.8  | 13.0  | 14.7  | 15–16       |
| Nagoya   | 3.9  | 5.1   | 5.9   | 6.5–7       |
| National | 94.3 | 104.6 | 117.0 | ~137        |

Source: Yamaguchi (Aug. 1984), p. 476.

**Table 4.3**  Infrastructure provision

| | Japan | US | W. Germany | France |
|---|---|---|---|---|
| Road paving (%) | 56 | 52 | 99 | 100 |
| Road/km² land area | 3.0 | 0.7 | 2.0 | 1.5 |
| Piped water (%) | 94 | 99 | 99 | 99 |
| Sewerage (%) | 34 | 72 | 91 | 65 |
| Flush toilets (%) | 58 | 99 | 94 | 79 |
| City parks (m²/person)* | 2 | 24 | 37 | 12 |

Note: * In Tokyo, Chicago, Bonn and Paris.
Source: Foreign Press Centre (1987), pp. 68, 86–7.

they bought properties in rural areas on the urban fringe, causing land prices in those areas to rise accordingly, and forcing people to move even further afield to buy affordable housing. The changes in metropolitan population growth rates are shown in Table 4.4: population growth was highest near the metropolitan centres before the 1960s, but since then the highest growth has occurred further and further away from the centres while population in the centres has grown more slowly or even fallen.

One offshoot of this move to the urban periphery has been the development of extensive urban sprawl which by now covers most of the

**Table 4.4**  Population changes in metropolitan areas

| City | Distance (km) | Growth rate (%) | | | | | Share (%) |
|---|---|---|---|---|---|---|---|
| | | 1955–60 | 1960–5 | 1965–70 | 1970–5 | 1975–80 | 1980 |
| Tokyo | 0–10 | 13 | −1 | −7 | −7 | −6 | 14 |
| | 10–20 | 30 | 25 | 12 | 6 | 2 | 30 |
| | 20–30 | 23 | 40 | 31 | 26 | 9 | 20 |
| | 30–40 | 15 | 37 | 44 | 30 | 14 | 22 |
| | 40–50 | 3 | 15 | 20 | 22 | 16 | 14 |
| | Total | 19 | 20 | 16 | 13 | 6 | 100 |
| Osaka | 0–10 | 21 | 12 | 2 | −3 | −4 | 29 |
| | 10–20 | 20 | 41 | 33 | 20 | 7 | 24 |
| | 20–30 | 13 | 21 | 25 | 22 | 8 | 15 |
| | 30–40 | 8 | 13 | 16 | 13 | 9 | 17 |
| | 40–50 | 2 | 5 | 5 | 7 | 3 | 15 |
| | Total | 14 | 17 | 13 | 9 | 4 | 100 |
| Nagoya | 0–10 | 19 | 14 | 6 | 3 | 0 | 27 |
| | 10–20 | 12 | 24 | 23 | 20 | 9 | 23 |
| | 20–30 | 8 | 14 | 19 | 16 | 11 | 18 |
| | 30–40 | 7 | 9 | 7 | 8 | 5 | 24 |
| | 40–50 | −1 | 1 | 3 | 7 | 5 | 8 |
| | Total | 11 | 13 | 11 | 10 | 5 | 100 |

Source: Nakai (1988), p. 200.

'Tokaido corridor' from Tokyo to Kobe. Other factors which contribute to this urban sprawl include:

— fragmented land ownership and small plot size;
— unwillingness of urban farmers to sell land, due both to the income which they derive from selling their products in the urban market, and to the fact that land price increases exceed interest rates;
— land readjustment, by which land is developed and exchanged without necessarily being put on the market;
— exemptions from some planning regulations for development projects below a certain size; and
— infrequent use of government expropriation powers.[5]

Another negative consequence of the move away from city centres has been that commute times have become very long. Transport costs, which in other countries limit the distance which people are willing to commute, are less of a consideration in Japan because companies often pay their employees' commuting costs. The average one-way commute to Tokyo is more than an hour, and 20 per cent of Tokyo's commuters travel for more than two hours each way. In the Osaka region, the average commute takes almost an hour, and 12 per cent commute more than two hours each way.[6] The strain on public transport is correspondingly great: in 1980 the ratio of passengers to carrying capacity in commuting trains during the busiest one-hour period was 210 per cent in Tokyo, 178 per cent in Osaka, and 200 per cent in Nagoya.[7]

Housing standards in Japan have improved steadily, as shown in Table 4.5. Between 1963 and 1983, average floor space increased by one-fifth and the number of rooms per house increased by a quarter, and are now comparable to those in European countries.[8] However, these

**Table 4.5**  Changes in housing

|  |  | Owner-occupied | Public | Rented Private | Company | Total |
|---|---|---|---|---|---|---|
| Tenure | 1963 | 64 | 24 | 15 | 7 | 100 |
| (%) | 1973 | 59 | 27 | 22 | 6 | 100 |
|  | 1983 | 62 | 25 | 22 | 5 | 100 |
| Floor | 1963 | 91 | 35 | 44 | 53 | 73 |
| space | 1973 | 103 | 36 | 40 | 54 | 77 |
| (m²) | 1983 | 112 | 39 | 42 | 57 | 86 |
| No. of | 1963 | 4.6 | 2.6 | 2.3 | 3.0 | 3.8 |
| rooms | 1973 | 5.2 | 2.8 | 2.4 | 3.2 | 4.2 |
|  | 1983 | 5.9 | 3.2 | 2.6 | 3.5 | 4.7 |

Source: Donnison and Hoshino (1988), pp. 194–5.

average figures conceal other underlying trends. The difference in size between owner-occupied and private rented accommodation is greater in Japan than in other countries. Most of the recent improvement in housing has been in the owner-occupied and public rented sector, while the standards of other types of accommodation have improved only marginally or, in the lower end of the private rented market, even decreased. Mere size is also not an indication of good living: the vast majority of Japanese houses have only a miniscule garden or none at all, and most have limited parking facilities. Much of the new housing is also in the form of high-rises, which not only are impersonal and crowded, but also rob much-needed sunlight from surrounding lower houses.

Several trends will influence future demand for land use and development in Japan. First, the proportion of aged people (65+) is expected to increase sharply, from 6 per cent in 1960 and 9 per cent in 1980 to 16 per cent in 2000 and 24 per cent in 2020.[9] This will increase demands for health care facilities and low-cost housing, especially since fewer elderly people are expected to live with younger relatives in the future. Second, Japan's general affluence, and the resulting increased orientation towards leisure rather than work, is expected to increase the demand for recreational facilities and park space. Similarly, as workers spend more time at home, better quality of housing will also be in demand.

## Planning framework

### Introduction

Japan's planning system is highly centralized, and policy and plan making is often a struggle between ministries with conflicting policy interests and directions. The Economic Planning Agency (EPA) and the National Land Agency (NLA) are responsible for drawing up plans for national development. These plans outline the policy (EPA) and physical (NLA) aspects of development and are derived through negotiations with other administrative bodies, as shown in the figure. The resulting plans do not involve a budget commitment; the NLA and EPA must coordinate this separately with the ministry/agency charged with carrying out the proposed plan.

Twenty-two ministries and agencies form the basis of executive power in Japan. Of these, eight are responsible for spatial policy: the MoC, MoT, MITI, MAFF, NLA, EA and the Okinawa and Hokkaido Development Agencies. The MoC is the most influential body concerned with city planning, and has considerable power over physical planning matters, including city planning, plan approval, land readjustment, urban redevelopment, housing, roads, urban parks, sewage systems and other urban facilities.[10] The MoT's responsibilities cover harbours, airports and railway development. Transport and land use planning interact closely, and at times there is intense rivalry between the MoC and MoT.[11] MITI controls industrial development and related location decisions. The MAFF is concerned with forest, marine and agricultural resources. In addition to its national planning role, the NLA is responsible for planning Japan's three metropolitan regions (NCR, Kinki, Chubu) and five of its non-metropolitan regions (Hokuriku, Chugoku, Shikoku, Kyushu, Tohoku). The EA is concerned with national parks and the development of regional environmental management plans (REMPs). The Okinawa and Hokkaido Development Agencies deal with the development of their respective areas.

At the local level, a two-tier government system, prefectural and municipal, operates as discussed in Chapter 2. The ten designated cities have the financial resources to carry out many public works without central government support. However, local autonomy is otherwise often undermined by the central government's financial controls.

## National development plans

Japan's national administration is oriented very much in favour of economic growth, and Japan's land and development plans are integrally linked to its economic plans. Figure 4.1 outlines Japan's planning system.

**Figure 4.1**   Summary of planning system

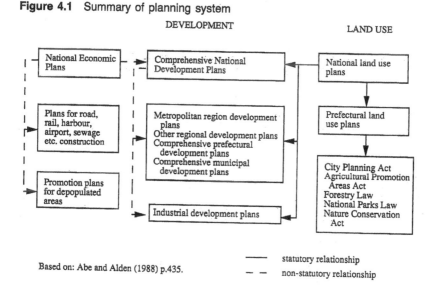

Based on: Abe and Alden (1988) p.435.

——— statutory relationship
— —  non-statutory relationship

Source: Based on Abe and Alden (1988), p. 435.

At the national level, a system of national economic plans, comprehensive national development plans and other long-term plans set out a basic approach to national land use. These plans feed down to the prefectural and municipal levels where they interact with and are interpreted by plans determined through legislation.

## National economic plans (NEPs)

The NEP of 1955 (Five-Year Plan for Economic Self-support) was the first post-occupation economic policy statement designed solely by Japanese planners. The plan emphasized the need to expand the petrochemical industry, and to foster the use of petroleum as the country's main energy source.[12] The realization of its goals after only two years signalled the end of Japan's postwar recovery period.[13]

Other plans followed rapidly as GNP growth rates consistently exceeded planned growth (see Table 4.6). During this period of rapid economic growth, development plans focused on providing infrastructure and on reducing regional disparities by relocating industries. For instance, the Ten-Year Income-Doubling Plan of 1960 aimed to link the industrial areas along the Pacific coast – Tokyo, Nagoya, Osaka and Kitakyushu – into a massive industrial belt. Several new industrial complexes were planned with the aim of realizing full employment and an annual growth rate of 7.2 per cent, reducing regional disparities, and improving living standards. The plan's announcement stimulated capital investment, and the national income was doubled two years ahead of schedule.

The plans were less successful, however, in increasing the standard of living. In particular, the location of industrial complexes near large population centres resulted in severe pollution problems and aggravated urban–rural differences.[14] The Economic and Social Development Plan of 1967 tried to deal with these and other problems[15] by aiming for stable

**Table 4.6** National economic plans

| Plan period | Plan title | Growth (%): Plan | Actual |
|---|---|---|---|
| 1955–60 | Five-Year Economic Self-Support Plan | 5.0 | 8.7 |
| 1958–62 | New Long-Range Economic Plan | 6.5 | 9.9 |
| 1961–70 | Ten-Year Income-Doubling Plan | 7.2 | 10.7 |
| 1964–8 | Medium-term Economic Plan | 8.1 | 10.6 |
| 1967–71 | Economic & Social Development Plan | 8.2 | 10.9 |
| 1970–5 | New Economic & Social Development Plan | 10.6 | 5.9 |
| 1973–7 | Basic Economic & Social Plan | 9.4 | 4.2 |
| 1976–80 | Economic Plan for the latter half of the 1970s | 6.0 | 5.1 |
| 1979–85 | Seven-Year Economic & Social Plan | 5.5 | 4.2 |
| 1983–90 | Economic & Social Guidelines for the 1980s | 4.0 | — |

rather than rapid growth and emphasizing social development.

Since 1970 planned GNP growth has not been achieved (see Table 4.6). Japan's economic plans now emphasize efficiency and cost-cutting, overseas investment, greater domestic spending, basic research and innovation, and the production of higher value-added products. Recent NEPs have focused on improving amenities and fostering the high-tech and information markets.

### Comprehensive national development plans (CNDP)

The need for a land development planning system was identified in the Comprehensive National Land Development Act of 1950. This act, which marked the beginning of Japan's national/regional planning system, set up an institutional framework for development planning which is still a basis of Japan's planning system. The system establishes development goals and broad principles for resource allocation and project promotion through the implementation of CNDPs. To date, four CNDPs have been prepared: by the EPA in 1962 and 1969, and by the NLA in 1977 and 1988.

The first CNDP (1962) aimed to balance economic growth, disperse industrial and urban development, and reduce regional inequalities in response to increasing pressures placed on the national government by rural authorities angered by the inequitable distribution of the benefits of modernization.[16] The plan was expansionist, proposing the creation of giant new industrial centres in remote regions to host heavy industrial and petrochemical complexes. These in turn were expected to stimulate further industrial development. This development, and rapid expansion of the nation's transport infrastructure, was expected to redistribute population and income.

Between 1964 and 1966 twenty-one industrial development areas were selected from more than forty applications.[17] However, this strategy proved to be ineffective, as metropolitan areas kept growing and regional disparities remained. In addition, local authorities spent about ¥600 million ($1.5 million) on lobbying alone, roughly equivalent to the total funds allocated by the national government for new infrastructure for the first year of the programme.[18]

The CNDP of 1969 added a spatial dimension to the 1967 NEP (Economic and Social Plan) by identifying seven planning regions and projecting their 1985 income and population. It aimed to disperse industrial development using a two-part strategy: (1) creation of a high-speed transport network of expressways and Shinkansen lines to link Japan's various regions, including the construction of bridge links between Honshu and Shikoku; and (2) the establishment of a small number of large industrial projects in areas of low economic growth. However, the development of industrial centres was again less than expected, due in part to the major pollution incidents and the slow rate of economic growth after 1973.[19]

One non-statutory but influential plan was prime minister Kakuei Tanaka's *Nihon Retto Kaizo-ron* (Remodelling the Japan Archipelago) of 1972. Like the earlier CNDPs, it assumed that regional disparities could be redressed by relocating industries to economically backward areas. It called for the development of a large number of regional cities with a population of about 250,000 located around an industrial nucleus, with a green belt to separate residential and recreational areas from the industries. The plan's publication sparked a rapid increase in speculative land purchases and land prices. However, public response was primarily hostile: the land speculation seemed primarily to benefit Tanaka's LDP party, the proposal for industrial centres was said to spread rather than contain pollution, and the proposed cities threatened to destroy the sense of community of the small rural villages which they would replace.[20]

The third CNDP of 1977, influenced by the recession and increased environmental consciousness of the time, radically departed from the predominantly expansionist thinking of the previous two plans. It emphasized a holistic and decentralized approach, and strengthening of local autonomy based on the promotion of provincial city regions with populations of 250,000–500,000. Several of these 'integrated residential zones' (*teijuken*) were to be established in each prefecture for a total of 200–300 nationwide, and within these zones the harmonious coexistence of the natural, living and industrial environments would be promoted. The plan emphasized the need to contain the size of the large metropolitan areas, and proposed that the capital and many government functions be moved away from Tokyo.[21]

The most recent CNDP (1988) is once again more development-oriented. It aims to stimulate the domestic economy and create an 'exchange network' of major information and transportation projects which will link various regions within Japan and internationally. Based on the assumption that the concentration of population and service industries in Tokyo is irreversible, the 1986 draft CNDP promoted Tokyo's development into a world-class centre of business, finance and information. However, this emphasis on the unilateral development of Tokyo was strongly opposed by the other cities and rural prefectures. The final plan proposes a multipolar national land structure, with regions specializing in particular types of development and linked by the 'exchange network'. The Kansai area is expected to become an academic and cultural centre, and a focus of finance and stock-market activities. The Nagoya area is projected to be a centre for research and development in advanced industrial technologies such as new ceramics and aerospace.[22]

*Regional development plans*

Metropolitan region development plans

Metropolitan region development plans for the National Capital Region (NCR), Kinki and Chubu areas (see Figure I.3) are based, respectively, on development acts of 1956, 1963 and 1966.[23] These plans have three levels – basic, consolidation and project – which cover about ten, five and one-year periods respectively. The basic plans provide a framework for population growth and land use, and set out policy for infrastructure provision. Consolidation plans determine improvements needed in public facilities and regional infrastructure. Project plans set out the projects needed to implement the consolidation plans. To date, four basic plans have been drawn up for the NCR and Kinki areas, and three for Chubu.[24]

The first NCR basic development plan of 1958 tried to redress some of Tokyo's problems of industrialization and urbanization by applying methods proposed in the Greater London Plan of 1944. It restricted construction in certain built-up areas. It designated satellite 'urbanization promotion areas' near Tokyo's 100-km radius, linked by radial roads and designed to absorb industrial expansion. It also proposed a British-style green belt around Tokyo to structure urban growth and provide agricultural and recreational space. However, this proved to be unfeasible because in the time it took to approve the plan urban sprawl had already virtually covered the proposed green belt.[25]

More recent metropolitan development plans take urban and suburban development for granted, and instead attempt to provide as much green space and infrastructure as possible. The emphasis of the basic plans for Kinki and Chubu is on integrating the regional structure, consolidating built-up areas and providing infrastructure. Kinki also aims to promote its historical and natural features while less-industrialized Chubu emphasizes industrial development.[26]

Long-term metropolitan development plans for the first quarter of the twenty-first century have been drawn up. The Plan for Remodelling of the National Capital Region proposes the creation of nine major satellite cities around the NCR to pull development pressure away from central Tokyo. The 'Subaru Plan' for the Kinki area and another plan for Greater Nagoya aim to revitalize the regions' manufacturing, industrial and commercial sectors.[27]

Other regional development plans

Development plans for Tohoku, Hokuriku, Chugoku, Shikoku and Kyushu (see Figure I.3) are formulated by the NLA in consultation with the prefectural governments. Hokkaido and Okinawa have their own land agencies to prepare development plans.

Comprehensive prefectural and municipal development plans

Comprehensive prefectural development plans have been prepared by all the prefectural governments and about 90 per cent of municipal governments. Often the plans are merely statements of intent rather than definite plans with budgets and projects ready for implementation. Moreover, they are optional and some prefectural authorities refrain from updating them so as to avoid putting undue pressure on the central government.[28] Instead, the prefectures adopt a more informal negotiation-oriented approach by preparing non-statutory documents. Prefectures also prefer to deal with individual ministries in relation to specific public works projects where they can be more certain of the availability of funding.

*Industrial development plans*

Industrial development plans promote industrial relocation through both positive and negative incentives, in an attempt to redress regional disparities and control over-concentration of people and industry in the cities. In keeping with the metropolitan region development acts, the NCR and Kinki region were divided into:

— existing urban zones (EUZs) where construction and expansion of industrial developments[29] is restricted;
— suburban consolidation zones (SCZs) where planned development, infrastructure provision and environmental conservation are promoted; and
— city development zones (CDZs) outside the urban centres where industrial development is encouraged so as to reduce pressure on the EUZs.

Chubu was divided into city consolidation zones (CCZs) and CDZs.

Various laws restrict the construction and expansion of industrial developments in EUZs,[30] promote consolidation of the SCZs/CCZs and development of the CDZs,[31] and promote the relocation of factories from 'Relocation Promotion Areas' in metropolitan areas to 'Inducement Areas' in remote low-growth regions.[32] Still other laws[33] assist local authorities in SCZs/CCZs and CDZs in providing housing, educational facilities, and infrastructure. However, these laws have not been particularly effective: few factories have relocated, in part because planners fear that their relocation will encourage urban decline. The suburbs are also facing agricultural decline due to the mixing of built-up and agricultural areas.[34]

To promote industrial development in outlying areas, twenty-one industrial cities and special industrial development areas were designated as a result of the first CNDP of 1962. Incentives such as tax relief and low-interest government loans were given to companies which moved to those areas. The central government gave special funding to local authorities, and gave priority to those areas for central government investments in transport and other infrastructure projects.

The most recent industrial/regional development plan is the 1983 'Technopolis Plan' which is expected to be implemented by 1990. The plan proposes the creation of regional centres composed of industrial, academic and residential facilities, located near a 'mother city' with a population of at least 150,000. The technopolises are expected to have a core of high-tech industrial centres specializing in next-generation technologies such as robotics, mechatronics, new materials or bio-technology; academic institutions oriented towards the development of this new technology to provide scientific/technical training and research; and housing and recreational facilities.

MITI first proposed the technopolis concept in 1980 as a demonstration project. The Technopolis Plan was drawn up in 1982, and the Law for Accelerated Regional Development Based on High-Technology Industries ('Technology Law') was enacted in 1983. MITI chose sites for technopolises based on local availability of high-tech industries, academic institutions with a high-tech orientation, enough land and water, and existing transport infrastructure. The government approved fourteen technopolis programmes in 1984, and five additional sites have since been approved.

The Technopolis Plan was expected to be primarily implemented by local governments and industries, especially small- and medium-scale firms. This strategy was expected to unlock the vitality of the private sector, and promote local initiatives and governmental/industrial/academic cooperation. Some central government aid was provided by MITI subsidies for frontier technology development and joint ventures, MoC construction of infrastructure, and NLA research on land and water in technopolis areas. However, most of the impetus for the plan has come, as planned, from host prefectures and local businesses. Eighteen of the nineteen technopolis sites have attracted substantial numbers of high-tech firms, and high-tech industries' investments in these areas have risen quickly.[35]

Future plans, established in 1986, call for the creation of twenty-eight research cores with the goal of promoting innovation and experimentation in small- and medium-size high-tech firms.[36] However, early experiences at Tsukuba Science City[37] have identified possible problems which may face the implementation of the technopolis and research core plans.[38]

*Land use plans*

National land use plans

The National Land Use Planning Act of 1974 is Japan's broadest policy for land use planning and control. It established a series of national, prefectural and local land use plans to ensure the coordinated development of national land resources, environmental conservation, and control of land use and values. The national plans are administered by the MoC and the NLA. The MoT and MAFF are responsible for plans related to transport and agriculture.

When the NLUPA was first enacted, control of land values was a priority, since corporate relocation, urbanization and the rampant land speculation caused by PM Tanaka's plan had resulted in urban sprawl and soaring land prices in the previous years. The NLUPA was expected to counter this by regulating land trading through two main provisions:

— the prefectural governor (or mayor of a designated city) can designate areas in which land prices have risen or are expected to rise rapidly, so that all land trading within those areas requires his/her permission; and
— the governor must be notified prior to the sale or purchase of land tracts over a given size,[39] and if he/she finds the price or intended use of the land inappropriate he/she may nullify or change the contract.[40]

These provisions proved to be successful in curbing the increase in land prices. In the year following the NLUPA's enactment, prices dropped by 9.2 per cent nationwide and stayed low for several years. Land prices rose again in 1980 when people's increased earnings led to an increased demand for housing, but remained stable thereafter due to tighter money policies and high land and construction costs. In 1987 land prices in Tokyo soared due to demand for office space, but rose only slowly in areas outside the large metropolitan centres.[41]

The NLUPA also sets basic policies for land use. The NLA prepares a National Land Use Plan for five land use categories: urban, agricultural, forest, natural parks and nature conservation. This is implemented in regional and local plans.

Prefectural land use plans

The National Land Use Plan is spatially interpreted in prefectural land use plans, which broadly indicate on maps the location of the five land use categories. These designations have no legal value and often overlap.[42]

Local land use plans

Local authorities are responsible for detailed land use control, which is implemented through the City Planning Act, Agricultural Promotion Areas Act and Nature Conservation Act. Urban land use in Japan was originally controlled by the City Planning Act of 1919. Although it tried to provide a reasonably comprehensive planning framework using land use zoning, the law was not designed to cope with the massive development that took place in the 1950s and 1960s, and provided neither standards for infrastructure nor land use controls for development.

The City Planning Act of 1968 provides a much more sophisticated and comprehensive system of land development control. The act is effective only in City Planning Areas (CPA), whose designation is decided about every five years by the prefectural governor in consultation with interest groups and with approval from the MoC. CPAs are either (1) existing built-up areas and their surroundings, or (2) areas currently not built up but which should be built up on planning grounds. CPAs cover about one quarter of Japan's land area, but due to the scarcity of habitable land about 90 per cent of the population lives in CPAs.

The governor divides the CPAs into urbanization promotion areas (UPAs) and urbanization control areas (UCAs). UPAs are built up or expected to be built up within ten years. Development in UPAs is generally permitted within the limits imposed by 'use zoning': UPAs are divided into eight primary use types,[43] and the use and volume of buildings within those zones are controlled.[44] Since it was amended in 1980, the City Planning Act has required comprehensive local plans to be drawn up for UPAs to promote forward planning: these must include, for example, provisions for natural disasters, water supply and public participation.

In contrast with UPAs, development in UCAs is quite strictly regulated. No development is allowed in principle, so as to control disorderly expansion of the urban area. Permission to build in UCAs is given only in exceptional circumstance.

However, due to Japan's difficult housing situation, these planning controls have been relaxed in recent years. UPAs have been extended, development controls within UCAs have been relaxed to permit more types of development, plot ratios in UPA zones have been increased and one-off relaxations of zone controls have become more frequent. For instance, since 1974 local authorities have been able to designate 'building lots' in UCAs in which development is freely permitted, so in practice UCAs are often sprinkled with developments despite their controlled status.[45] Unfortunately, these policies have done little to lower land prices and control urban sprawl, and it has been suggested that they actually aggravate the situation.[46]

Urban land use is further controlled by various laws which, for example, preserve green spaces, cultural assets and agricultural land on urban fringes.[47]

### Agricultural land use plans

Agricultural land use was originally controlled by the MAFF's 1959 Standards for Permitting the Conversion of Farmland. However, these did not control urban sprawl. The Agricultural Promotion Areas Act (1969) set up a national system of zoning for agricultural land. Within APAs, permission from the governor or the MAFF is required before farmland can be used for other functions. The Reserved Agricultural Land Act (1974) establishes farmland as a city planning zone for environmental and safety purposes, and as possible reserved space for public utilities.

### *Other plans*

Transport infrastructure is controlled by a series of MoT plans for the construction of roads, railroads and harbours starting in the late 1950s, and five-year plans for airport development starting in 1966.[48] Allocations for public utilities are determined in conjunction with the economic plans. MoC sewerage construction is funded by a series of five-year plans starting in 1963. Other plans give special aid to depressed areas such as mountain villages, outlying islands, coal mining areas and heavy snow areas.

### *Effectiveness of planning system*

Japan's development and land use plans have had several consistent objectives: to prevent over-concentration of people and industries in the large metropolitan areas; to reduce regional disparities; and to provide infrastructure, green space, and amenities, and generally raise the living standard. These objectives have clearly not been met.

The populations of the three largest metropolitan areas have consistently risen faster than the national average, and at present account for more than 40 per cent of the country's total population. The three metropolises produce nearly half of the nation's employment and more than half of its income,[49] and about two-thirds of new factories are still being located there.[50]

Regional income differences, shown in Table 4.7, decreased from a peak in the early 1960s to their lowest level in 1977. Since then they have risen again slightly, due in part to declining public investment and changes in the industrial structure after the oil crises.[51]

As shown in Table 4.3, Japan's provision of infrastructure, especially

**Table 4.7**    Regional income differences (national average = 100)

| Region | 1960 | 1970 | 1980 |
|---|---|---|---|
| Capital | 127 | 121 | 115 |
| Kinki | 115 | 112 | 104 |
| Chubu | 101 | 100 | 97 |
| Hokuriku | 92 | 87 | 93 |
| Hokkaido | 89 | 83 | 93 |
| Chugoku | 83 | 91 | 93 |
| Shikoku | 79 | 83 | 84 |
| Kyushu | 73 | 73 | 86 |
| Tohoku | 73 | 73 | 83 |
| Okinawa | 56 | 51 | 67 |
| Max. difference | 71 | 70 | 48 |

Source: Abe and Alden (1988), p. 432.

park space, lags far behind that of other highly industrialized countries, and behind its own goals.[52] Housing is also a problem, with 57 per cent of houses under the Average Housing Standards and 14.5 per cent under the Minimum Housing Standard in 1980.[53] More than 45 per cent of surveyed households were dissatisfied with their dwelling, although 70 per cent were satisfied with their surroundings.[54]

A major reason for the planning system's difficulty in achieving its objectives is its top-down approach, with economic goals at the top. Any subsequent plans at best can only mitigate the negative side-effects of the rapid development required to attain economic goals.

These constraints also limit the effectiveness of environmental policies. Japan's planning framework is extremely rigid with approximately 300 kinds of plans.[55] The planning process can be divided into four basic stages: vision plan, master plan, programme plan and project plan. At present, EIA is introduced only at the final project stage, where most of the details have already been decided and a budget has, in many cases, been allocated. At such a late stage environmental considerations can lead to only superficial changes and have little influence on the final outcome of the project. Environmental management and EIA play a minor role in Japan's planning system, and environmental factors are considered only when they infringe on primary goals.

### Environmental management plans

Regional environmental management plans (REMPs) have been developed to complement the EIA system in an effort to overcome these problems. REMPs are non-statutory plans which originate at the local authority level (the first REMP was established in September 1973 by

Osaka prefecture) and which promote the systematic implementation of environmental protection measures. They state future goals for environmental quality and the specific programmes needed to achieve those goals, and invite participation from the local community.[56] The plans have the following characteristics:

— timescale of ten to fifteen years;
— integration/coordination of various efforts for pollution prevention, nature protection and the creation of environmental amenities;
— establishment of agreed-on goals for all members of the community (local authorities, residents and developers); and
— encouragement of voluntary activities.

If the REMPs are well coordinated with existing mandatory plans, they can fulfil the local land use plans' requirements for environmental consideration. The REMPs can also serve as a basis for judgements made by the local authorities on environmental issues when they formulate future development plans. As of March 1986, thirty-one prefectures and designated cities had implemented some form of REMP scheme.[57] Table 4.8 lists major cities' REMP schemes.

Osaka prefecture's system exemplifies the relationship between REMP and EIA. The REMP establishes environmental carrying capacities, and from these the maximum permissible release of air and water pollutants is calculated for each area ($500km^2$ on average). Through computer simulations, pollutant concentrations are regularly calculated for all locations. The impacts of individual projects can then be checked against the REMP to ensure that environmental capacities are not exceeded. However, experience with the Kansai International Airport (Chapter 11) shows that the system, although admirable, is neither fully developed nor beyond abuse.

**Table 4.8**  Cities with REMP systems

| City | Name of plan | Established |
| --- | --- | --- |
| Kobe | Environmental Management Plan for Air and Water | May 1974 |
| Kawasaki | Regional Environmental Management Plan | July 1977 |
| Osaka | Basic Plan for Protection of Waterways | May 1983 |
| | Basic Plan for Protection of the Air | Jan. 1984 |
| Yokohama | Environmental Management Plan | Mar. 1986 |
| Kyoto | Environmental Management Plan | Apr. 1986 |
| Kita-Kyushu | Environmental Management Plan | Apr. 1986 |
| Fukuoka | Environmental Plan | Sep. 1986 |

Source: Hattori (1987), p. 4.

## Conclusions

Japan's planning system is very complicated, sophisticated and rigid. Economic growth, development and technology are main features of the plans. Those agencies responsible for development, particularly the MoC, have acquired much power and political momentum. Once plans are produced, their proposals are actively pursued and implemented despite possible objections. Japan's postwar planning has been a great success for those concerned with promoting rapid development and attainment of international respect; Japan has planned, developed, consumed and produced its way to its status as an economic superpower.

On the other hand, environmental management in Japan is in its preliminary stages. The lack of a legal basis for environmental management is compounded by the weak standing of the bodies responsible for environmental protection. Environmental policy-making is affected by several agencies' economic and land use plans. This makes coordination and integration difficult, and cripples efforts to incorporate environmental priorities into the administrative structure.

Environmental legislation to date has sought only to remedy the negative side-effects of development through pollution control laws. The concept of sustainable development has gained little support. Preventive legislation, like an EIA law, which seeks to ensure that the environment is not harmed by proposed developments, has been difficult to introduce. The next chapter discusses environmental policy in Japan.

## Notes

1. Nakai (1988), pp. 198–9.
2. The rural population was 52 million in 1950 and 33 million in 1960. The urban population was 31.2 million in 1950 and 60.4 million in 1960: Tokyo Metropolitan Government (1971), p. 26.
3. In late 1987, for instance, office rental in Tokyo cost more than twice as much as that in London, and three times that in New York: Foreign Press Centre (Oct. 1987), p. 29.
4. Nakai (1988), p. 202.
5. Zetter (1986), p. 202.
6. Foreign Press Centre (1987), p. 83.
7. Ministry of Transport (1985), p. 8.
8. Japan's average floor space of $87m^2$ (in 1983) compares with 80 in W. Germany, 77 in France, and 140 in the US (in 1978/9). Japan's average 4.7 rooms/house compare with W. Germany's 4.5, France's 3.7 and the US's 5.1: Donnison and Hoshino (1988), p. 195; Foreign Press Centre (1987), p. 86.
9. Ibid.
10. Siman (1989), pp. 13–16.
11. Ibid.
12. Coal and hydropower were the main pre-war energy sources.
13. Huddle and Reich (1975), p. 84.

14. Ibid., p. 93.
15. E.g. internationalization of the economy, labour shortages.
16. Gottman (1980).
17. Fifteen industrial city areas were designated under the Industrial City Development Act of 1962, and six industrial development areas were created under the Act for the Promotion of Industrial Development of Special Areas of 1964.
18. Huddle and Reich (1975), p. 93.
19. Alden (1984), pp. 64–6; Yamaguchi (1984), pp. 476–8; Abe and Alden (1988) p. 434.
20. Ibid.; Tamura (1987), p. 379.
21. Alden (1984), pp. 64–6; Hebbert (1986), p. 145.
22. Ibid. The Kansai area is the area around Osaka.
23. The National Capital Region Development Act, the Kinki Region Development Act and the Chubu Region Development Act.
24. Yamaguchi (1984), pp. 479–80. Plans for the NCR were released in 1958, 1968, 1976 and 1987; for the Kinki area in 1965, 1971, 1978 and 1988; and for Chubu in 1968, 1978 and 1988.
25. Alden (1984), p. 72.
26. Yamaguchi (1984), pp. 480–2.
27. Ibid., p. 482.
28. Siman (1989), pp. 13–16.
29. Industrial developments include academic facilities.
30. The 1959 Act Concerning the Restriction of Industry and Other Functions in Existing Urban Zones of the National Capital Region, and the 1964 Act Concerning the Restriction of Industry and Other Facilities in Existing Urban Zones of the Kinki Region establish zones in the bounds of the EUZs, and within these zones new construction and expansion of industrial and research facilities is permitted only in exceptional circumstances. By March 1982, 267 permits had been given in the NCR, and 66 in the Kinki Region under these laws.
31. The 1958 Act Concerning the Development of the Suburban Consolidation Zone and City Development Zones of the National Capital Region, and the 1964 Act Concerning the Restriction of Industry and Other Facilities in Existing Urban Zones of the Kinki Region instituted programmes of land development for industrial parks. The Special Taxation Measures Act gives tax preference to firms which give up business assets outside an industrial park in exchange for assets inside a park. By May 1983, 24 industrial parks covering about 5,800ha were located in the NCR, and four parks covering about 1,400ha were in the Kinki Region. The laws also promote the construction of roads, drainage facilities and other infrastructure.
32. The 1972 Act for the Promotion of Industrial Relocation designates zones in the NCR, Kinki, and Chubu regions – the Zones for Restriction of Industry etc. (see note 30) for the NCR and Kinki (excluding parts of Kyoto and Kobe), and old built-up areas of Nagoya – as Relocation Promotion Areas (RPA). Areas in prefectures with low population growth and industrialization are designated as Inducement Areas (IA). Various financial incentives are given to induce factories to relocate from RPAs to IAs.
33. The 1966 Act Concerning Special Measures Pertaining to State Financing of the Consolidation of the Suburban Consolidation Zones and Similar Zones in the National Capital Region, Kinki Region and Chubu Region set up fiscal measures to promote projects undertaken under the consolidation plans for

the SCZs and CDZs of the NCR and the Kinki region, and the CCZ and CDZ of the Chubu region (starting in 1969 for Chubu, 1966 for the other areas). The measures authorized the prefectures to issue more local government bonds for the provision of housing and infrastructure, and to subsidize the payment of interest on these bonds. They also raised the percentage of local authority infrastructure construction costs borne by the central government.

Since 1973, local authorities have been able to tax agricultural land within a UPA at its residential value, and since 1974 they could apply a capital gains tax after short-term land holding. Since 1975 the local authorities could also collect a 'workshop tax' based on the volume of business conducted at existing offices and workshops; and on the floor surface, workshop use, and wages at newly constructed or expanded buildings.

34. Foreign Press Centre (Oct. 1987), p. 9.
35. The technopolis concept is discussed further in, e.g., Kawashima and Stoehr (1988), Fujita (1988) and Glasmeier (1988).
36. The research cores would have four functions: (1) to act as research centres for cooperative industry/academic/government use; (2) to train researchers; (3) to provide information, communication and exhibition facilities; and (4) to provide space and business/management services.
37. Tsukuba Science City is a research and academic city which lies about 60km northeast of Tokyo. It comprises about 30,000ha and has a planned population of about 200,000. Plans for Tsukuba were first proposed in 1963, construction began in 1966, and in 1970 the Act for Construction of the Tsukuba Academic New Town was passed. By 1984, 45 research and academic facilities, including many central government research institutes, were located in Tsukuba.
38. Some of these problems are: (1) Difficulty in attracting workers away from their current jobs to the remote technopolis sites. Many workers live in temporary accommodation at the technopolis during the week, and return to their families in the cities on weekends; (2) The combination of high-tech industries and academic facilities does not necessarily lead to improved or accelerated research and development, and the establishment of high-tech industries does not necessarily boost local high-tech development; and (3) The relatively weak level of central government funding and the designation of a rather large number of technopolises may dilute the programme into unproductive competition for limited funds.
39. 2,000m$^2$ in UPAs, 5,000m$^2$ in CPAs, and 10,000m$^2$ elsewhere. In 1987 the NLUPA was amended so that in Tokyo notification must be given for transactions involving 500m$^2$ or more.
40. Foreign Press Centre (Oct. 1987), pp. 14–15.
41. Ibid., pp. 2–3.
42. Hebbert (1986), p. 145.
43. Exclusive residential categories I and II, residential, neighbourhood commercial, commercial, quasi-industrial, industrial, exclusive industrial.
44. Building volume is limited by the permitted plot ratio (total floor area to site area). Planning permission is given to all development schemes which conform to the area's use and plot ratio. No permission is needed for developments of less than 1,000m$^2$ in zoned UPAs or 3,000m$^2$ in unzoned UPAs, developments by the central or local government or public corporations, or developments related to land readjustment.
45. Foreign Press Centre (Oct. 1987), pp. 8–9.

46. Relaxations in land use zoning may exacerbate rather than improve the situation, since land availability is restricted by the unwillingness of landowners to sell their land, not by planning controls. Relaxation of zoning controls could thus actually increase land prices, since it increases the development value and thus the price of land. Thus the present policies are expected to result in urban sprawl and inefficient land use, and zoning controls should be tightened: Nakai (1988), pp. 197–216.
47. These include the Urban Space Conservation Act (1973), which designates historic green space conservation areas in which tree cutting and building alterations are controlled; the Special Law for the Preservation of the Historical Landscape (1966), which sets strict controls to maintain landscape as it was when it was designated; the Urban Parks Improvement (Emergency Measures) Law (1972); and the Urban Fringe Agricultural Areas Improvement Act (1968). Preservation Zones in the Kinki and Chubu regions preserve green space, cultural assets and areas for tourism. Suburban Green Area Preservation Zones in the NCR and Kinki region help to prevent pollution and provide areas of safety in the case of natural disasters.
48. Japan Transport Economic Research Centre (1982), p. 30a.
49. Yamaguchi (1984), pp. 475–6; Abe and Alden (1988), p. 432.
50. Foreign Press Centre (Oct. 1987), p. 35.
51. Abe and Alden (1988), pp. 432–3.
52. Alden (1984), p. 133.
53. Zetter (1986), p. 136. Average housing standards are $50m^2$ for two persons, $86m^2$ for four; minimum housing standards are $29m^2$ for two persons, $50m^2$ for four.
54. Foreign Press Centre (1987), p. 86.
55. Morita (1981).
56. Environment Agency (Apr. 1986), p. 1.
57. Nomura et al. (1985), p. 1.

# Chapter five

# Environmental policy

## Introduction

As international understanding of Japan's influence on the global environment has grown, Japan's environmental policies and the institutions responsible for their implementation have been subject to increasing scrutiny and pressure. In response, the members of a Cabinet meeting on the global environment agreed on 30 June 1989 that Japan should take a lead in creating an international framework for environmental protection. In July 1989 the administration, headed by PM Sousuke Uno, stated that 1989–90 would be Japan's year of global environmental diplomacy.[1] Although we should be grateful that Japan recognizes the seriousness of the environmental problem, we should also cautiously consider whether it is in a position to take a lead on environmental issues.

A complex set of regulations, standards and practices make up a nation's environmental policy. Although many similarities exist between policies, each nation's policy has distinctive elements which result from factors such as its legal structure, economic system and cultural context.[2] Japan's system of pollution control is a valuable example for other countries. However, Japan also has some lessons to learn from other countries, especially with regard to nature conservation and public participation.

This chapter examines the scope and effectiveness of environmental policy in Japan. It begins with a brief discussion of the institutions involved in environmental policy-making, and of the form and history of Japan's environmental legislation. However, such a limited examination may be misleading because legislation may not be implemented as intended. Therefore the chapter concludes with a discussion of some of the mechanisms used to implement the policy, and the levels of expenditure on environment-related matters.

**Policy-makers and implementation**

The structure and form of Japan's environmental policy-making has not changed significantly since 1971 when the EA was established. The EA was expected to transform the decentralized and *ad hoc* nature of environmental concerns into more formal and centralized procedures and institutions.[3] Unfortunately, as discussed in Chapter 2, the EA is not the sole agency concerned with environmental matters. For instance, it has no jurisdiction over

— aircraft noise;
— discharge of hazardous substances;
— generation, storage and disposal of radioactive wastes;
— pollution control in specified factories;
— special government financial measures for pollution control;
— settlement of pollution disputes; or
— pollution crimes relating to human health;

and has only an advisory role on

— regulation of agricultural chemicals;
— waste disposal and sewerage; and
— marine pollution.

These concerns are dealt with by MITI, the MoC, the MHW and the MAFF.

As mentioned in Chapter 4, the EA also has no control over land use and development plans. Land use controls fall under the jurisdiction of the MoC and the NLA, both of which are pro-development and unwilling to relinquish their power. Although the EA and the NLA are located in the same building, they rarely communicate and they have no joint committees. The Nature Conservation Act (1972) and the National Land Use Planning Act (1974) have been used to control development, but are generally ineffective against the pressures favouring development.

The EA's attempts to overcome jurisdictional rivalries and promote an integrated environmental policy have been continually hampered by ministerial interference and vested interests. The major ministries placed many of their own members in key positions in the EA: these members remained loyal to their original ministries, undermining the EA's autonomy. In fact, the EA reminds one of Cinderella, bullied by her ugly step-sisters (the other ministries) and her step-mother (the LDP), banished to sweep cinders in the kitchen, and dreaming that one day her prince (widespread environmental awareness) will come.

The local authorities, however, have played a leading role in developing environmental policy. Under the 1947 amendments to the Local Autonomy Law, local authorities can enact legislation unless such legislation violates a national law. The environmental standards set by local governments in the 1950s and 1960s were at times more stringent than those of the national government. They also emphasized the total volume of discharges, whereas government regulations focused on emission standards and ambient air and water quality. This prompted a flurry of debates in the late 1960s on the legality of local autonomy. The outcome of this was that local authorities were given greater power over environmental matters than before and were permitted to establish stricter standards than those of the national government.[4]

In turn, the national government was affected by the local authority policies. For instance in the late 1960s, in response to the central government's weak NO$x$ vehicle emission standards, seven large cities promoted strict emission standards and banned cars in particularly polluted districts; this prompted the national government to strengthen its own standards.[5] Yokkaichi city's system of compensation for pollution-related health damages (enacted in 1965) was a model for the national system of 1973. Local authority EIA guidelines (from 1973) and regulations (from 1976) also preceded and influenced the national EIA guidelines of 1984.

However, the powers of local authorities are limited. An elaborate consensus has to be worked out before environmental laws can be implemented, first between the EA, the other ministries and the LDP, and then between the local government, the polluter and the regional administrators. This complicates and delays the implementation of legislation and policy-making.

The main barrier to implementing comprehensive environmental policies in Japan is the lack of a strong centralizing force at the national level. Even the budget allocation system encourages fiscal decentralization and undermines administrative unity in environmental affairs; this will be discussed later in this chapter.

### National and local environmental laws

As mentioned in Chapter 3, Japan has instituted one of the world's most comprehensive environmental legislative systems. This system is underpinned by the Basic Law For Environmental Pollution Control (1967). Other major national laws concerned with the control of specific types of pollution are:

Air pollution
  Air Pollution Control Law (1968)
  Road Transport and Motor Vehicle Law (1951)
  Road Traffic Law (1960)
  Electric Power Industry Law (1964)
  Gas Industry Law (1954)

Water pollution
  Water Pollution Control Law (1970)
  Sewerage Law (1958)
  River Law (1964)
  Marine Pollution Control Law (1970)
  Hazardous Substances Control Law (1950)
  Agricultural Soil Pollution Prevention Law (1970)
  Seto Inland Sea Environment Conservation Law (1973)
  Regulations for waste treatment and disposal

Soil pollution
  Agricultural Soil Pollution Prevention Law (1970)

Noise and vibration
  Noise Regulation Law (1968)
  Vibration Regulation Law (1976)
  Road Transport and Motor Vehicle Law (1951)
  Road Traffic Law (1960)

Ground subsidence
  Industrial Water Law (1956)
  Law Concerning the Pumping of Groundwater for Use in Buildings
  (1956)

Offensive odour
  Offensive Odour Control Law (1971).

Other environment-related laws include:

— The Nature Conservation Act (1972) which set up an administrative
  framework for nature conservation, including surveys of the natural
  environment and the establishment of wilderness and nature conservation areas;
— The Pollution-Related Health Damage Compensation Law (1973)
  which established a system which charges polluting facilities and
  automobile owners based on the amount of pollution they generate,
  and distributes the funds to certified pollution victims; and
— The Chemical Substances Control Law (1973) which requires
  manufacturers and importers of new chemicals to submit information
  on the chemical's properties to the MHW and MITI, prohibits the sale
  of chemicals until their safety (persistency, cumulative tendency,
  toxicity to humans) is proven, and strictly regulates chemicals which
  do not meet safety criteria but have no harmless substitute.

Appendix B provides a more comprehensive list. The enactment of this
broad array of environmental laws illustrates Japan's ability to adapt
quickly once organized and motivated. By the mid-1970s, Japan had one
of the most complete statutory frameworks for environmental policy and

some of the strictest environmental standards in the world. The laws emphasize compensation and dispute resolution and in many cases were enacted before their counterparts in the West.[6]

Local governments' environmental ordinances are a vital part of Japan's environmental policy. In particular, the Tokyo Metropolitan Government was a forerunner in enacting such ordinances, with regulations for the control of industrial pollution (1949), noise (1954), smoke and soot (1955) and environmental pollution (1969). The last of these aimed to prevent and eliminate all pollution that could disrupt the quality of life of Tokyo citizens, and was unique in that it stressed environmental issues over the need for economic growth.[7] Osaka also enacted a pollution prevention ordinance in 1964, and initiated the preparation of regional environmental management plans (see Chapter 4) in 1973. By 1975, every prefecture had passed some form of pollution control ordinance.[8] In 1987, the number of municipal governments which had enacted ordinances for pollution control, environmental conservation, and matters related to pollution control were, respectively, 496, 310 and 1,390.[9]

The Japanese approach to environmental policy-making has, however, overlooked or misinterpreted several elements found in environmental policies elsewhere:

— It lacks awareness of ecological issues, with potentially serious implications for nature conservation. The Basic Law defines pollution as air, water and soil pollution, noise, vibration, land subsidence and offensive odours. The definition omits radioactivity. It also only attaches significance to known pollution causes which affect human health and the living environment.[10] The act does not address ecosystem degradation unless it affects human health, and as such ignores the fact that human existence depends on the stability of the ecosystem.[11] Some local authorities have adopted a much wider definition: for instance the Tokyo Metropolitan Government defines pollution as 'any infringement on the environment', which can also apply to yet-unknown forms of pollution.[12]
— Japan's environmental laws emphasize remedial pollution abatement measures rather than anticipatory or preventive measures. An example of this is the EA's continued failure to introduce an EIA law due to inter-ministerial conflict and opposition from economic circles.
— Public participation in policy-making is extremely limited: it is generally restricted to letters of complaint, opinion surveys and public hearings. Hearings, when they take place, tend to be strictly regulated explanatory sessions rather than forums for meaningful participation. Despite continued calls by citizens and lawyers for the introduction of a freedom of information act like that of the US, the national

government is unlikely to expand public participation. Instead, it is promoting the sense that the administration can be entrusted to control pollution problems; most citizens are all too willing to accept this, and this has undermined the power of Japan's environmental protest groups.[13]

— Japan has failed to adequately consider global environmental issues and has been reluctant to control environmentally destructive activities of Japanese companies overseas. To date, environmental laws have not been interpreted to apply outside Japan and administrative guidance has not covered external environmental issues.[14]

## Mechanisms for policy implementation

Japan's mechanisms for environmental protection are shown in Figure 5.1. Their primary emphasis is on the control of pollution through the application of strict environmental standards and the use of advanced pollution control technology.[15] The implementation of this system is supported by legislative enforcement and administrative guidance. Anticipatory policies such as nature conservation, EIA and REMPs are relatively undeveloped. This section discusses mechanisms for environmental protection, including environmental standards, regional pollution control programmes, pollution control agreements, compensation, nature conservation, administrative guidance and environmental monitoring.

### Environmental standards

Environmental standards are a basic tool for environmental pollution control. There are two kinds of standards in Japan: emission standards and environmental quality standards (EQS). The former are legally binding limits on the concentration of pollutants in the effluent from a

**Figure 5.1** Mechanisms for environmental protection

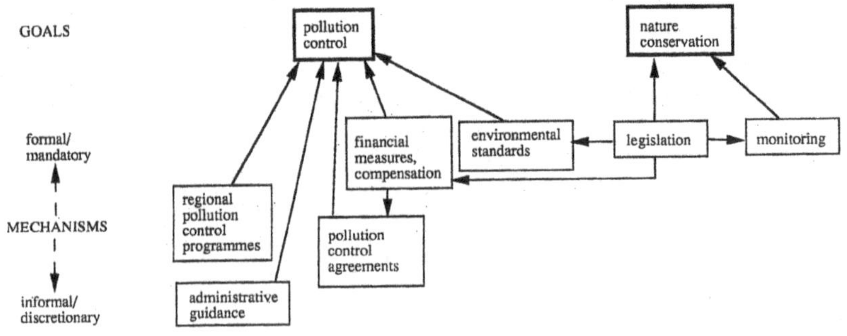

given source. The latter are policy objectives for the levels of pollution in a given area.[16] Emission standards are set by both the central government EA[17] and local authorities. In many cases the prefectural and municipal authorities have established much stricter and a wider range of standards than those of national administration. EQSs are also set by the EA, but more consultation is necessary.[18] Appendix A lists Japan's standards for various pollutant types. Japan's standards are stricter than those in most other countries,[19] and they are generally rigorously enforced.

Two EQSs pertain to water quality. The first relates to substances which affect human health (e.g. cyanide, lead, PCBs) and is uniformly applied to all water bodies subject to public use. The second is concerned with environmental indicators (e.g. pH, COD, suspended solids) and depends on the type of water body (river, lake or coastal) and its planned use.[20] Effluent standards related to human health and environmental protection have also been set, and are less stringent than the EQSs.

Ambient air quality standards have been established for $SO_2$, CO, $NO_2$, photochemical oxidants and particulates. Ambient readings should not exceed these standards over a fixed period. Emission standards cover $SO_2$, soot and dust, $NOx$, and harmful substances (e.g. chlorine and lead compounds). Separate exhaust standards exist for different types of stationary facilities and motor vehicles. Permissible levels for noise (including specific standards for aircraft and Shinkansen), and offensive odours have been set.

In particularly polluted areas, the total amount of emissions is also controlled. A 1974 amendment to the Air Pollution Control Law established area-wide total emission controls for $SO_2$; to date, twenty-four regions covering about half of the country's total fuel consumption are designated as total emission control regions. A similar system for $NOx$ was set up in 1981 and applies to Tokyo, Yokohama and Osaka; in these cities industrial activity is so concentrated that it is difficult to attain EQS solely with emission control.[21]

However, several problems still exist. First, except in already severely polluted areas there are no provisions to control the total amount of pollution discharged. Second, only the concentration of pollutants at the outlet is considered. This allows a factory to emit the original amount of pollutants while conforming with the standards by simply diluting its waste-water. Ambient standards do little to reduce the impact of large quantities of pollutants annually settling into the ecosystem.

Since the early 1970s, the EA has been studying the concept of 'environmental capacity', which considers how much waste an ecosystem can handle before its powers of self-purification break down. If this concept were applied, the construction of new developments would only be permitted if the additional pollution caused by their construction and operation did not bring the area's total pollutant levels above the carrying capacity. This concept has not, as yet, been implemented.[22]

### Regional pollution control programmes

Regional pollution control programmes (RPCP) for particularly polluted areas were set up under article 19 of the Basic Law with the aim of facilitating the attainment of EQSs. In RPCP areas, the local authorities are required to establish comprehensive measures for environmental pollution control. The programmes are financed by the national and local government. The first areas, Yokkaichi and Mizushima, were designated in 1970. At present, forty-one areas, covering 9 per cent of the national land area and 54 per cent of the population, are designated as RPCPs. In 1986, approximately ¥1.35 trillion were allocated for control projects in these areas.[23]

Except in controlling $SO_2$ levels, these programmes have not been impressive. As shown in Table 5.1, even with the high levels of funding available EQS compliance rates have not increased significantly. The RPCPs have been criticized for their over-dependence on central government support which in turn has resulted in programmatic uniformity and subservience to industrial interests. The programmes have concentrated on short-term measures and neglected long-range issues such as land use and urban design. They have also placed too much emphasis on attaining a limited number of EQSs to the detriment of wider environmental concerns.[24]

### Pollution control agreements

Pollution control agreements between local authorities and industries define mutually acceptable control measures for noise, odours, air and water pollution. The first agreements began in the mid-1960s as a result of local governments' increasing autonomy and industries' weakened bargaining power. Early agreements were rather abstract, but more recent examples are much more stringent, mandating strict emission

**Table 5.1**   EQS compliance rates in RPCPs (%)

|  | 1980 | 1981 | 1982 | 1983 | 1984 | 1985 | 1986 | 1987 |
|---|---|---|---|---|---|---|---|---|
| $NO_2$ (<6ppm) | 94 | 95 | 97 | 98 | 95 | 98 | 96 | 91 |
| Particulates | 19 | 25 | 32 | 56 | 39 | 40 | 47 | 41 |
| BOD | 57 | 54 | 58 | 55 | 53 | 56 | 58 | 55 |
| COD | 27 | 27 | 13 | 24 | 24 | 18 | 24 | 35 |

Source: Environment Agency, various publications.

standards, use of low-sulphur fuels, and advanced pollution prevention technologies. Some agreements also specify strong enforcement measures like liability for damage, fines, cancellation of contract or even interruption of the municipal water supply. By the late 1980s, 1,600–2,000 pollution control agreements were drawn up annually.[25]

These agreements often involve local citizen groups and are seen as part of the democratization process of corporate/government decisions: for example, in 1985–6 residents were signatories in twenty-six local government/industry agreements, and observers at another sixty. Moreover, 229 agreements were concluded between residents and industries directly.[26] Many agreements entitle the public to inspect factories and obtain information related to pollution control issues.

One major problem with these agreements, however, relates to their legal validity; scholars believe that local authorities do not have the legal powers to impose fines or other penalties by contract.[27] It is also uncertain whether the agreements are enforced.

### Compensation system

The pollution-related health damage compensation system was until recently one of the most notable features of the Japanese approach to environmental pollution. The system has its origins at the local authority level, with measures introduced by Yokkaichi (1965), Nanyo-cho (1967) and Takaoka (1968) to cover the medical costs of victims of pollution-related diseases. The first national system was set up in December 1969 by the Law for Special Measures for the Relief of Pollution-Related Health Damage. This system was like the local systems in that it provided for medical costs rather than compensation, and relied on public monies and contributions from enterprises. In April 1972, the EA began to consider how to improve the system. At this time, the verdicts for the 'Big Four' pollution trials were coming through, which legally recognized the plaintiffs' claims for both consolation money and compensation for loss of income. The government hurriedly drafted a bill based on recommendations by the Central Council on Environmental Pollution Control (CCEPC), which had been studying the problem at the EA's request. The Pollution-Related Health Damage Compensation Law was enacted in October 1973.[28]

According to the law, compensation must be paid for two types of pollution-related diseases. Class I areas are those in which a statistically significant correlation of pollution and particular diseases (e.g. chronic bronchitis, asthma) has been identified. Class II areas are those in which diseases have been related to specific pollutants (e.g. cadmium or mercury poisoning). Persons living in these areas and suffering from the specified diseases had to apply to the local authorities to be certified as pollution victims; upon certification they were entitled to compensation for medical costs, disability payments, and in the case of death 'condolence payment'.[29] Table 2.1 gives an indication of the number of certified pollution victims and outstanding applications.

Although far from perfect,[30] Japan's compensation system had significant political and social repercussion. It represented a positive step towards the integration of environmental and social costs into economic development. It had no parallel elsewhere in the world.

However, as discussed in Chapter 2, under pressure from industry the EA's director consulted the CCEPC on the future of class I areas. The CCEPC reported in October 1986 that no new patients should be certified. One year later the law was amended, and it became effective in March 1988.[31]

*Nature conservation*

The Nature Conservation Act of 1972 gave the EA responsibility for planning and promoting basic policies for nature conservation, and for coordinating the conservation-related activities of other agencies. Article 5 of the law also requires the EA to undertake surveys of the natural environment (the 'National Green Census'). These surveys cover such items as the distribution of flora and fauna, habitats and growth of important plant communities, and areas of natural scenic beauty. Three surveys have been carried out to date: in 1973, 1978–9, and 1983–7. The fourth began in 1988 and is as yet unfinished.

In order to conserve the natural environment the EA can designate conservation areas. As of March 1989, five areas covering 5,631ha had been designated as wilderness areas, and nine areas covering 7,550ha were nature conservation areas. The prefectural authorities also have powers under local ordinances to designate nature conservation areas: in March 1989, 501 such areas covering 72,092ha existed.[32]

Under the Natural Parks Law, the EA can designate natural parks. In Japan, there are three types of parks; national, quasi-national and prefectural. In 1987 the combined area of these parks covered roughly 14 per cent of the national land area: twenty-eight (2 million ha) were national parks, fifty-four (1.3 million ha) were quasi-national parks, and about 300 (2 million ha) were prefectural parks. In addition, there were respectively 27, 30 and 57 national, quasi-national and prefectural marine parks. Within the parks, areas of special scenic beauty have been designated: in these areas, the EA's director or the prefectural governor must be consulted prior to the construction of new buildings or the expansion of existing activities. However, these special protection areas represent only 6 per cent of the total park area.[33]

Although Japan has implemented a wide range of measures to ensure the conservation of the natural environment, many parts of the national parks are being extensively developed. The nature conservation areas are threatened with logging, conversion of primary forest to plantations, rural revitalization, river management projects (e.g. dams, concrete lining), and shoreline conversion.[34] Only five of the national parks are recognized by the International Union for the Conservation of Nature as national parks using criteria established by the United Nations List of National Parks and Equivalent Reserves. Seven more are recognized by the IUCN as Scientific/Nature Reserves. These areas combined cover only 59,335ha, roughly 0.015 per cent of Japan's total area.[35]

The passage of the 1988 Resort Law could also severely affect the natural environment. The law could allow 39,700km$^2$ (roughly 11 per cent of the national land area) to be developed as resort areas. Much of this development pressure would fall on the less well-protected national park areas.[36]

### Administrative guidance

The enforcement of pollution control policies relies heavily on administrative guidance, which may involve loans, grants, subsidies, tax incentives and licence approvals. The use of administrative guidance explains the dissociation that exists between legal imperatives and actual behaviour. Even when industry commits a serious illegality under a pollution control law, the government usually responds by negotiating, exhorting and giving constructive suggestions, but seldom by seriously threatening to prosecute.[37]

### Environmental monitoring

The control and use of information is a very powerful tool. Environmental monitoring can help to monitor changes in environmental quality, provide a basis for communication between decision-makers and the public, evaluate the effectiveness of environmental policies, and identify the need for new policies.

Over the past thirty years, a wide network of stations to collect and process environmental data has been established in Japan, based on various pollution control laws.[38] This network includes more than 1,600 stations to monitor air pollution, 300 for automobile exhaust and 270 for water quality. The Japanese government spends about ¥15 billion annually on the collection and analysis of environmental information.[39] The resulting data have been particularly important in policy formulation and the functioning of REMPs and EIA systems.

## Expenditure

The level of funding allocated for pollution control and environmental conservation reflects its relative importance in terms of national priorities. In 1962 only 0.2 per cent of Japan's GNP was allocated for pollution control. When environmental conditions worsened this rose to 1–2 per cent,[40] and by 1988 anti-pollution spending represented about 1.9 per cent of the ¥336.5 trillion GNP as follows:

| | |
|---|---|
| Ministries | ¥1.3 trillion |
| Associated institutions | ¥1.4 trillion |
| Local authorities | ¥3.4 trillion |
| Industry | approx. ¥0.3 trillion |

### *National government expenditure*

As shown in Figure 5.2, government spending for environmental matters rose by about 25 per cent annually in the 1970s, from ¥83 billion in 1970 to ¥1.2 trillion (2.4 per cent of the national budget) in 1982.[41] After 1982 these expenditures decreased due to several factors:

— All aspects of the administration were subject to a financial squeeze.
— In December 1981, the Keidanren recommended that the balance between environmental and economic priorities should be reconsidered, and that environmental policy should be rationalized.
— In 1982 a series of reforms began under the administrative reform committee. One sub-committee recommended the abolition of both EIA and the compensation system. Although this proposal was not implemented, the prospects for Japan's environmental policy were not promising from that time on. The selection of Yasuhiro Nakasone as PM in November 1982 gave further impetus to the reform programme.

Environmental spending rose again after 1987, and stood at ¥1.34 trillion (2.3 per cent of the national budget) in 1990. The increased spending should, however, not be confused with the re-emergence of environmental priorities. Spending by the Ministry of Finance, which accounted for 16 per cent of the 1988 environment-related budget, only began to be counted in 1988; it comes from the sale of the national telecommunications firm and is used as a temporary loan for the construction of sewage works. The increase also reflects spending on noise prevention related to the national construction boom.

Approximately 85 per cent of the government's environment-related budget is allocated to pollution prevention projects: of this about 70 per cent is for sewerage development, and much of the rest is for noise abatement around airports.[42] About 10 per cent is allocated to nature conservation, and the rest is used to promote research and development,

**Figure 5.2**   National government environment-related expenditure (¥1bn)

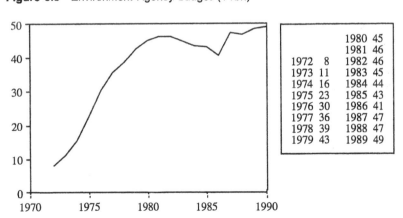

| 1970 | 83 | 1980 | 1166 |
|------|------|------|------|
| 1971 | 112 | 1981 | 1205 |
| 1972 | 169 | 1982 | 1192 |
| 1973 | 274 | 1983 | 1177 |
| 1974 | 342 | 1984 | 1147 |
| 1975 | 375 | 1985 | 1117 |
| 1976 | 485 | 1986 | 1094 |
| 1977 | 627 | 1987 | 1087 |
| 1978 | 868 | 1988 | 1284 |
| 1979 | 1125 | 1989 | 1329 |
|      |      | 1990 | 1340 |

Expenditure by ministry/agency (1988)

- MoC
- MoF
- Defense
- MoT
- MHW
- Hokkaido Devel. Agency
- EA
- Other

Expenditure by subject (1988)

- Pollution prevention
- Nature conservation
- Research & development
- Compensating pollution victims
- Other

Source: Environment Agency, various publications.

**Figure 5.3**   Environment Agency budget (¥1bn)

|      |      | 1980 | 45 |
|------|------|------|------|
|      |      | 1981 | 46 |
| 1972 | 8 | 1982 | 46 |
| 1973 | 11 | 1983 | 45 |
| 1974 | 16 | 1984 | 44 |
| 1975 | 23 | 1985 | 43 |
| 1976 | 30 | 1986 | 41 |
| 1977 | 36 | 1987 | 47 |
| 1978 | 39 | 1988 | 47 |
| 1979 | 43 | 1989 | 49 |

Source: Environment Agency, various publications.

compensate pollution victims, monitor and enforce environmental offences, and establish standards.

The amount of government funding allocated to the EA decreased steadily from 1983 to 1986, as can be seen in Figure 5.3. This reflects the severe restraints on spending at all government levels rather than a specific attempt to limit environmental improvements. However, it is a disturbing trend when one considers the cost of damage caused by pollution each year, and the environmental degradation which is likely to result from the government's planned construction projects. Moreover, increasing calls for the EA to adopt a more global stance on environmental issues means that ever greater requirements will be made on a tighter budget. This will make the EA increasingly vulnerable to external pressure to ease pollution control policies.

The government promotes pollution control and environmental conservation through several other methods:

— It allocates approximately ¥1.4 trillion annually from the Financial Investment and Loan Scheme to subsidize the pollution control activities of several organizations. The scheme invests government funds and the deposits from pensions and postal saving accounts. Chief among the recipient organizations are the Japan Development Bank and the Pollution Control Service Corp.[43]
— It encourages companies to install pollution control equipment through tax incentives which provide time-limited tax favours and depreciation on pollution control equipment. These effectively reduce the cost of the equipment. For example, in 1976 first-year depreciation write-offs of 50 per cent and 33 per cent were permitted for, respectively, new pollution control equipment and non-polluting equipment. Similar policies are used by local authorities.[44]
— The Law on Special Financial Arrangements for Regional Pollution Control (1971) helps local authorities to finance pollution control projects such as the construction of monitoring stations, sewage systems, waste treatment plants, green buffer zones, and river and harbour dredging. The law includes provisions for national government subsidies and increases in the national government's share of pollution control expenses.

The Pollution Control Works Cost Allocation Law (1970) provides financial support for conservation measures (see Chapter 3).[45]

*Local authority expenditure*

Local authority environmental concerns include monitoring and research, sewerage and waste disposal developments, and relief for pollution

**Figure 5.4** Local authority environment-related expenditure (¥1bn)

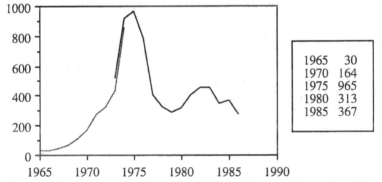

| | | | |
|---|---|---|---|
| 1971 | 370 | 1981 | 2900 |
| 1972 | 810 | 1982 | 2840 |
| 1973 | 950 | 1983 | 2760 |
| 1974 | 1200 | 1984 | 2670 |
| 1975 | 1400 | 1985 | 2760 |
| 1976 | 1510 | 1986 | 2970 |
| 1977 | 1880 | 1987 | 3410 |
| 1978 | 2280 | 1988 | 3400 |
| 1979 | 2520 | 1989 | 3396 |
| 1980 | 2750 | | |

Source: Environment Agency, various publications.

**Figure 5.5** Industry environment-related expenditure (¥1bn)

| | |
|---|---|
| 1965 | 30 |
| 1970 | 164 |
| 1975 | 965 |
| 1980 | 313 |
| 1985 | 367 |

Sources: Dotted line – Trade and Industry Technological Data Research Company (1975); plain line – Environment Agency, various publications.

related health damages. Local authority expenditure on environmental conservation, like that of the national government, rose rapidly in the 1970s, from ¥0.4 trillion in 1971 to almost ¥2.8 trillion in 1980 (see Figure 5.4). By 1988 it had risen to ¥3.4 trillion, approximately 6 per cent of local authority expenditure. The number of local authority employees working on environmental matters more than doubled between 1971 and 1976, and has risen further since.[46]

### Industry expenditure

Private firms also invest considerably in pollution control. As shown in Figure 5.5, large corporations spent a total of almost ¥5 trillion on pollution control in the 1970s, and about ¥4 trillion in the 1980s. The peak was in 1975, when firms spent nearly 18 per cent of their total capital expenditure on pollution abatement. Since then expenditure has decreased, partly because firms have largely completed their anti-pollution investment, and partly because the need to be seen to invest in pollution control has diminished as pollution levels have dropped.

An interesting feature of Japan's approach to pollution control is its wholehearted adoption of the Polluter Pays Principle, an economic concept introduced by the Organization for Economic Cooperation and Development in 1972. The PPP specifies that pollution control costs should be borne by the polluter, and thus the consumer through increased prices, rather than by the government/taxpayer. Japan has extended the original concept to include the restoration of polluted environments, administration of monitoring programmes and compensation for pollution victims. For example, all major $SO_2$ emitters must pay a fee based on the volume of $SO_2$ they release, even if their emissions comply with emission standards. The proceeds of this are used to reimburse patients with certified respiratory ailments and to research these illnesses.[47]

### Direction of Japan's environmental policy

The EA was created to ensure the formulation of an integrated environmental policy. But Japan's environmental policy to date has been concerned with the protection of human life and health and with the abatement of damage caused by environmental pollution. The EA's inability to broaden the scope of environmental policy[48] has led it to become over-reliant on technological aspects of pollution control. The EA feels that it plays a major role in fostering the development of environment-related technologies and disseminating known technology.[49] However, Japan's administration cannot reasonably hope for presently unknown, unimagined technological breakthroughs to solve all their problems.[50]

Of greater concern is Japan's wish to export this approach to developing nations.[51] The approach presupposes that only money and technology are needed to solve global environmental problems. Although the progressive financing and 'technology-forcing' legislation which Japan has implemented so successfully are essential to solving global environment problems, a more fundamental change in values is also needed to attain sustainable development.

However, within its present societal framework, Japan is not in a position to make these changes for several reasons. First, in Japan (and elsewhere) no consensus exists about the nature of an environmentally sound economy, nor does any real blueprint for an alternative approach to decision-making. The environment has historically been viewed from different perspectives by the government, industry and environmentalists.

Second, there is no authentic political leadership on environmental issues, and political activity is seldom tuned in to the public interest. To date, the ruling party has equated the public interest with economic growth. Environmental issues have been distorted by political suspicions and recriminations. Environmentalism is often confused with radicalism, and the EA is not strong enough to shift the balance of power towards greater environmental awareness.

Third, the government's lofty rhetoric about environmental problems has not been translated into working policies. Already in 1972, the CCEPC's planning committee recommended that Japan should move away from retroactive policy-making and towards anticipative and preventive policies. The committee also recommended the following points:

— environmental preservation should be incorporated into the objectives of economic and other policies;
— economic growth should be allowed only within the limits of policies designed to ensure effective environmental preservation;
— the industrial structure and consumption patterns should be modified to use fewer environmental resources;
— expansion of social capital related to the living environment (e.g. sewerage, parks) should be encouraged; and
— more resources should be invested in the development of pollution control technology.[52]

However, almost a decade later the committee admitted that they had failed to spread the view that environmental problems are attributable to interactions between humans and the environment. Their report of 1981 noted that environmental policy has progressed little since 1972. It made almost exactly the same recommendations as the first report, adding only a few new concepts:

— Optimal environmental use should be determined in economic and social terms. Environmental resources should be 'enhanced' by, for example, internalizing the social costs of using environmental resources into economic activities, and carrying out pre-development impact studies.
— A national consensus should be reached on the establishment of new comprehensive, anticipatory environmental policies; the current lack of such a consensus hinders the formulation and implementation of environmental policy.[53]

Of course, some improvements have occurred, particularly in relation to pollution research and monitoring. The local governments have been particularly active in promoting EIA and REMP systems, and in expanding public participation. However, the essential problem remains that neither the EA nor local governments have the power to forbid environmentally destructive projects from being built in relatively clean areas, nor their further concentration into already polluted areas.

Environmental policy in Japan has successfully reduced the scale of the environmental problem. This reflects a high level of commitment and cooperation by the administration and industry. However, only the rate of environmental deterioration has been reduced; the problem itself has not been solved. Environmental policy in Japan deals only with pollution as it affects human health and its related environment. Attempts to broaden the scope and effectiveness of environmental policy have been hindered by the inability to form a consensus at the national level. The history of the development of EIA in Japan illustrates the problems that the process of improving environmental policy faces.

### Notes

1. *Japan Times* (1 July 1989), p. 1.
2. Reich (1983), p. 191.
3. Ibid., p. 193.
4. In 1970 the Basic Law was amended to allow local authorities to establish stricter emission standards than those of the national government, in 1974 the Local Autonomy Law was amended to allow local authorities to control pollution, and other laws (e.g. the Air and Water Pollution Control Acts) were revised to expand local authorities' rights to enact environmental ordinances and enforce pollution control. The decisions to allow greater local government control over environmental policy were based on article 94 of the Constitution, article 14 of the Local Autonomy Law and local authorities' mandate to protect the health of their residents.
5. Weidner (1986), p. 50.
6. For example the Chemical Substances Control Law was passed three years before the US's equivalent Toxic Substances Control Act of 1976.
7. The Tokyo Metropolitan Environmental Pollution Control Ordinance consolidated and replaced Tokyo's three previous environmental ordinances.
8. Gresser *et al.* (1981), pp. 246–7.
9. Environment Agency (1989), *Quality of the Environment in Japan 1988*, p. 348.
10. Article 2 of the Basic Law for Pollution Control.
11. Kelley *et al.* (1976), p. 244.
12. McKean (1981), p. 222.
13. Morishima (1981), pp. 77–84.
14. Forrest (1986), p. 4.
15. Weidner (1986), p. 44.
16. Gresser *et al.* (1981), p. 154.
17. When setting standards, with the exception of those standards relating to automobile exhausts, the EA does not need to consult the CCEPC. However, the EA may seek technical advice from expert groups, particularly on feasibility issues, and it may collect outside opinions through direct consultations with interested parties: Gresser *et al.* (1981), p. 257.

18. The EA must consult the CCEPC when it sets EQSs. The CCEPC sets up sub-committees and expert committees to consider the standards. When reviewing the sub-committees' findings the CCEPC must provide industry, government agencies and environmental groups with the opportunity to communicate their views to the council members.

19. For example, the table below compares several countries' 1986 automobile exhaust standards (in g/km):

|  | CO | Hydrocarbons | NOx |
|---|---|---|---|
| Australia | 9.3 | 0.93 | 1.93 |
| Switzerland | 24.2 | 2.10 | 1.90 |
| US | 2.1 | 0.26 | 0.63 |
| Japan | 2.1 | 0.25 | 0.25 |

Noise control standards for automobiles in 1986 were (in dB) Australia 81, Switzerland 75, US 80, and Japan 78: Foreign Press Centre (1987), p. 69.

20. Different EQSs may be set for different rivers, or within the same body of water.

21. Environment Agency (1987), *Quality of the Environment in Japan 1986*, pp. 90–3.

22. Kelley *et. al.* (1976).

23. Environment Agency (1987), pp. 48–54.

24. Gresser *et al.* (1981), p. 262.

25. Between Oct. 1986 and Sep. 1987, 1,652 agreements were made; 1,994 were made the previous year, and 1,804 were made in 1984–5.

26. Ibid., p. 349.

27. Gresser *et al.* (1981), p. 249.

28. Tsukatani (1989), pp. 1–3.

29. Weidner (1986), pp. 53–6.

30. In particular, gaining certification as a pollution victim took a very long time.

31. Tsukatani (1989), pp. 22–3.

32. Environment Agency (1989, in Japanese) *White Paper on the Environment*, pp. 327–41. Translated into English and published annually as 'The Quality of the Environment in Japan'.

33. Ibid; Environment Agency (1988), *Environmental Quality in Japan 1987*, pp. 203–12.

34. By 1984, 54 per cent of Japan's shoreline had been altered by road construction, breakwaters, seawalls and major reclamation projects.

35. Suzuki (Sep. 1989), pp. 26.

36. Ibid., pp. 26–8.

37. For instance, between 1981 and 1985 on average almost 54,000 pollution-related complaints were filed with the police annually; of these about 1 per cent resulted in arrests, the rest in guidance or warnings. During the same period, only 6,140 pollution-related arrests (including arrests not instigated by public complaints) were made annually, 87 per cent of which were for incorrect waste disposal: Environment Agency (1987), p. 283.

38. E.g. the Air and Water Pollution Control Laws, the Law concerning the Prevention of Marine Pollution and Maritime Disasters.

39. Hashimoto, Z. (1985), p. 37.

40. The percentage varies depending on assumptions about management and operation costs, opportunity cost of capital, depreciation expenses or enterprises capitalized at less than ¥50 million, for which no figures exist: Kelley *et al.* (1976), p. 261; Corwin (1980), p. 154.

41. Hashimoto, M. (1985), p. 43.
42. Ibid., p. 44.
43. In 1987, the JDB received ¥71 billion, the EPCSC received ¥40 billion, and other institutions received a total of ¥4.4 billion.
44. Small firms were offered either 50 per cent depreciation on equipment costs in the first year or 30 per cent for each of the first three years. This rate was reduced to 25 per cent in 1977: Corwin (1980), pp. 154–7; Hashimoto, M. (1985), p. 47.
45. Hashimoto, M. (1985), pp. 45–6.
46. Prefectural environment-related staff rose from 2,634 in 1971 to 8,652 in 1985. Municipal staff rose from 3,411 in 1971 to 6,268 in 1985.
47. Corwin (1980).
48. E.g. the EA's failure to introduce EIA legislation or a law for lake conservation.
49. Environment Agency (1987), p. 3.
50. Kelley *et al.* (1976), p. 283.
51. Saburo Okita, Chairman, Ad Hoc Group on Global Environment Problems, stated recently that Japan, with its money and technological know-how, is in a very good position to assume a leadership role in global environmental protection. *Journal of Japanese Trade and Industry* (Oct. 1989), p. 11.
52. Environment Agency (1972).
53. Environment Agency (Mar. 1981), p. 1.

# Part two

# Environmental impact assessment

# Chapter six

# Development of environmental impact assessment

## Introduction

Over the last twenty years, EIA in Japan first progressed rapidly then slowly regressed, paralleling the rise and fall of Japan's environmental policy. The first rudimentary EIAs were performed in the late 1960s. In the early 1970s the US National Environmental Policy Act was used as a model for an EIA bill. Several laws were amended to include EIA provisions and a number of ministries incorporated EIA into their planning procedures. However, as experience with impact assessment grew it became obvious that the NEPA could not be adopted without threatening Japan's whole-hearted pursuit of economic growth. The pro-development ministries feared that EIA would result in legal battles, added costs and delayed development. The hierarchical nature of Japan's planning system required environmental factors to be considered with minimum changes to existing power structures and priorities. In 1984, after eight years of negotiations and revisions, the EA abandoned its efforts to introduce EIA legislation. Instead, the Cabinet approved a (non-enforceable) decision which incorporates the bill's main points.

Four local authorities have enacted (enforceable) EIA ordinances, and another twenty-three have adopted EIA guidelines. Generally, these cover a wider range of topics and have more stringent procedures than the national guidelines.

No changes are presently envisaged for this system. The Cabinet decision seems to have appeased public demand for comprehensive preventive environmental planning, and the country is once again focusing its energies on economic growth.

The following two sections review, respectively, the development of the national and local authority EIA systems. Table 6.1 summarizes this development, and Appendix C gives the texts of major EIA-related reports. The final section discusses why the EA failed to introduce an effective, comprehensive EIA system. The system's procedures are discussed in Chapter 7.

**Table 6.1**  Brief chronology of Japan's EIA system

1965  MITI performed pre-construction environmental investigations.

1972  Cabinet decision 'Environmental Conservation Measures Relating to Public Works'.

1973  Fukuoka prefecture established first local government EIA guideline.
Port and Harbour Law and Public Water Areas Reclamation Law amended to include consideration of EIA.

1974  CCEPC report 'Guidelines for Conducting EIA (Interim Report)'.

1975  Komeito and Socialist parties proposed EIA bills.
CCEPC report 'Concerning the Direction of the EIA System'.

1976  EA began negotiating with other ministries concerning EIA bill.
Kawasaki enacted first local government EIA ordinance.

1977  EA compiled technical index for EIA and guidelines for Honshu–Shikoku Bridge Project EIA.
MITI guidelines 'Strengthening of EIA and Environmental Review for the Siting of Power Plants'.

1978  MoC guidelines 'Policy for Interim Measures Concerning EIA for Public Works under the Jurisdiction of MoC'.

1979  MoT guidelines 'Implementation of EIA for the Construction of Five Bullet Train Lines'.
CCEPC report 'Recommendations on a System of EIA'.
Natural Resource and Energy Agency guidelines 'Implementation of Environmental Surveys and Environmental Review for the Siting of Power Plants'.

1981  LDP committees discussed problems of legislating EIA.
Power stations removed from projects subject to EIA. LDP and Cabinet agreed to bill. Bill presented to Diet.

1983  Diet dissolved; EIA bill nullified.

1984  Cabinet decision 'On Implementation of EIA'.
Committee for Fostering the Implementation of EISs report 'Common Matters Necessary for the Procedure etc. Based on Implementation Scheme for EIA'.
EA guidelines 'Principles Concerning Surveys and Studies, Prediction and Evaluation of Environmental Impact'.

## National EIA system

In early 1964, under pressure from citizens who opposed the proposed development of a *kombinato* near Mishima, MITI and MHW commissioned several scientists to research methods for reducing pollutants from the development. At the same time, the town's progressive mayor, Taiso Hasegawa, set up a group to investigate the development's environmental implications. The group's report was completed in May 1964, several months before the government study. It concluded that the development would cause air and water pollution, and has been described as Japan's first EIA.[1] These findings and the results of an opinion poll conducted shortly thereafter prompted the mayor to oppose the plan.

Shocked by this decision, and hoping to avoid further pollution incidents like that at Yokkaichi, MITI began conducting EIAs in areas where industrial development was planned from 1965 onwards. These early EIAs were limited to an analysis of the expected effects of SOx, COD, BOD, and oil pollution.[2] The EA's advice was sought for some of these projects, but seems to have been ineffective: for instance, the EA's guidance on the Tomakomai industrial complex in Hokkaido in 1972–3 was criticized for approving industrialization in an area where air pollution levels already exceeded ambient standards.[3]

The real beginning of EIA in Japan was the June 1972 cabinet decision 'Environmental Preservation Measures for Public Works'. This required central government agencies and ministries to conduct, for all public works,

> as necessary, prior surveys and investigations of the nature and degree of environmental impact, measures to prevent environmental degradation and the relative merits of alternatives, and [to] direct firms to take whatever steps are necessary in accordance with the findings of these surveys.[4]

Local governments were called upon to take similar steps.

Essentially, this document mirrored the sentiments expressed in the NEPA. However, it did not include public participation procedures. Furthermore, a Cabinet decision is only a form of administrative guidance and thus not legally binding. These factors prevented the courts from interpreting and expanding the document's provisions,[5] and gave the national agencies great freedom to decide what form their EIA procedures should take.

In December 1973 the newly established Agency of Natural Resources and Energy within MITI began requiring electric power companies to investigate the environmental impact of their proposed developments. MITI used this information when making its final decision on sites for power stations. In July 1977 MITI resolved to improve the power companies' environmental surveys and MITI's review of these surveys, and in June 1979 the agency issued detailed guidelines for implementing the 1977 resolution.[6]

In July 1978 the MoC released interim guidelines for performing EIAs on projects under its jurisdiction, and in January 1979 the MoT began requiring the Japanese National Railways and the Japan Railway Construction Corporation to conduct assessments of the environmental impact of new express train lines.

To conform to the Cabinet resolution, the Diet also amended the Port and Harbour Law in July 1973, the Public Waters Reclamation Law in September 1973 and the Factory Location Law in October 1973 to include provisions for EIA. The 1973 Interim Law for the Conservation of the

Environment of the Seto Inland Sea also included EIA provisions and was the first legislation to mention the phrase 'environmental impact assessment'.

In 1973–4 the EA began formulating guidelines for impact assessment. This was perhaps the time of greatest optimism for those involved in the development of EIA in Japan. The government was still unaware of the full consequences that the introduction of an EIA system would have on its policy-making process, and citizen movements were calling for the government to prepare a comprehensive statutory EIA system. In these early days, the EA planned to introduce EIA into the project plan stage of the administrative planning system (see Chapter 4), and hoped to later incorporate it into the programme stage, where a greater variety of alternatives could be considered.

In June 1974 the impact assessment section of the CCEPC released 'Guidelines for Conducting EIA (Interim Report)'. This report called for environmental assessment at all stages of planning and for the recognition that the assessment process is inherently uncertain and consequently should be subject to constant monitoring and feedback. A month later, the EA was reorganized and an environmental impact assessment division was set up to oversee EIAs; later its operation was expanded to include the drafting and implementation of EIA guidelines.[7]

The opposition parties also sought to optimize on the public concern about EIA. In January 1975 the Komeito Party submitted a bill to the Diet which would require certain industries to prepare comprehensive assessments of the environmental and social implications of their developments; these assessments would be reviewed by an administrative commission and rejected if unsound; the public would be informed of the proposals, could comment on draft assessments and could challenge the decisions of the commission; and the committee's decision could be subject to judicial review. In October 1975 the Socialist Party introduced a bill containing detailed procedures for the preparation of EISs, but without the permit procedures of the Komeito bill. The Japan Bar Association submitted a proposal similar to that of the Komeito Party, but broadened it to include assessment of new legislation and administrative policies.[8]

A national EIA bill has to be approved by several levels of government before being enacted:

— the large ministries, e.g. MITI, MoC, MoT, the Ministry of Finance;
— the Cabinet;
— the LDP's General Affairs Committee;
— the Diet's Environment Committee; and
— the Diet.

Although legislation lacks the flexibility of guidelines, it can be enforced, and violations can be made punishable by fines and/or public announcement of the violation.[9] Legislation also provides a strong mandate to developers to ensure that their EIAs are adequate, and gives a legal basis for environmental and citizen groups to challenge inadequate EIAs. Although the bills proposed by the opposition parties had little or no chance of being approved by the LDP-controlled Diet, they prompted the EA to begin formulating its own legislation.

In December 1975 the CCEPC released a draft outline of an EIA law, 'Legal System of Environmental Impact Assessment'. Based on this, the EA proposed a draft EIA bill in early 1976. The EA bill was much weaker than the others presented: it required an EIA only in cases which could have substantial environmental impact and only after a development plan had already been formalized; limited public participation by making public hearings discretionary; had no provisions for an independent environment committee to assess EIAs; and did not allow for judicial review.

At the beginning of 1976, the EA's bill was included in the Cabinet schedule for that year, and in June the ministries concerned with EIA – the EA, NLA, MITI, MoC, and MoT – met to discuss the bill. However, the best opportunity to introduce a comprehensive national EIA system had essentially already been lost. Economic changes, decreases in pollutant levels and decreased public interest made the prospect of legislating EIA more remote. The bill became subject to increasing criticism from the industrial sector, which argued that an EIA law would be used by militant protest groups to delay development and that, rather than resolving conflict, the law could potentially engender it. It was also argued that EIA techniques were underdeveloped and that introduction of EIA was premature.[10] In addition, as more and more ministries and local authorities developed their own EIA systems (see next section), the EIA process became sectionalized and increasingly difficult to unify.

These arguments were reinforced when the EA's bill was tested on the Kojima–Sakaide Route of the Honshu–Shikoku Bridge Project (Chapter 10). The EA commented harshly on the draft EIA of March 1978, prompting the Bridge Authority to claim that EIA would inevitably lead to delay and higher costs. As a result, other ministries, hitherto sympathetic to the EA's goals, turned against the idea of EIA legislation.

To counter these objections, the EA modified its bill several times. In March 1977 the bill's procedures for public participation were cut down by restricting public comment to those residents living in the affected area, and by limiting the conditions under which public hearings and briefings could be held.[11] The link between the EIA bill and future lawsuits was also severed as much as possible. In 1978 and 1979 the bill's emphasis was changed from 'reform' to 'information exchange', basically making EIA an addendum to the planning process rather than an integral part of it.[12] The EA's control over the EIA process was also weakened; the EA would decide only basic items for EIA procedures, but the ministries would be responsible for concrete guidelines; and instead of sending his opinion directly to the industry, the EA's director would send it to the concerned ministry, which would include it in its own report.[13]

In April 1979, after three years of study, the CCEPC made recommendations to the EA on a legal framework for EIA in a report entitled 'Council Recommendations on a System of EIA'. The report urged that legislation for EIA be produced at the earliest possible date, and prompted the EA to make its last major effort to introduce EIA legislation. In March 1980, fourteen ministries formed a Cabinet subcommittee to discuss the proposal. Two months later, after the need to consider alternatives was removed from the bill at MITI's insistence, the committee agreed on the bill.

In February 1981, the LDP's General Affairs Committee held several informal meetings and agreed to the bill after power stations were removed from the list of projects for which EIA must be performed.[14] In late April 1981 the bill was presented to the Diet. The Diet's Environment Committee met several times in 1982 and 1983 to discuss the bill and hear expert opinion. However, the time for deliberation was insufficient, and the bill became void due to the dissolution of the House of Representatives in November 1983.

The EA continued to work on the bill in early 1984, with the strong support of most local authorities. Arrangements were made to resubmit the bill to the Diet committee. In the meantime, the LDP was also considering the bill. The General Affairs Committee brought together five LDP committees (environment, commerce and industry, transport, construction, and local affairs) four times during that summer. At one of these meetings, opinions of the heads of local governments were heard, and at another opinions of concerned business groups were heard. The results of these deliberations were presented to the Cabinet on 8 August. They noted that EIA legislation was necessary, but that a number of problems remained:

— integration with existing regulations;
— the possibility that lawsuits would increase, delaying development; and
— the scope and form of public participation.

They concluded that the problem of developing suitable EIA legislation should continue to be studied, but that the bill should not yet be resubmitted to the Diet.[15]

Two days after this presentation, recognizing the futility of expending the EA's energies on an unpassable bill, the EA's director recommended to the Cabinet that the bill should be used as a basis for the development of national EIA guidelines. Such guidelines, he reasoned, would help to unify the local governments' EIA procedures. On 28 August 1984 the Cabinet passed a decision 'Implementation of Environmental Impact Assessment'.[16] This decision is further discussed in Chapter 7.

As part of the decision, a committee was set up to foster the implementation of the scheme and decide on matters not clarified in the decision itself. In November of that year the committee released guidelines which included:

— principles for determining the project undertaker and the competent ministry;
— recipients of EISs;
— interim measures; and
— procedures for industries undertaking more than one project subject to EIA.[17]

Six days later, on 27 November, the EA's director released 'Principles concerning Surveys, Studies, Prediction and Evaluation of Environmental Impact'.[18] The report provided details of matters which the ministries should take into account when formulating EIA guidelines. Since then, the ministries have issued guidelines, as shown in Table 6.2.

**Table 6.2** Ministry EIA guidelines

| Ministry | Project type | Date estab. | oper. |
|---|---|---|---|
| MoC | Dams, drainage projects | Sep. 1985 | Mar. 1986 |
| | Lake & marsh development | * | * |
| | Roads | Sep. 1985 | Mar. 1986 |
| | Land redevelopment | Mar. 1987 | Sep. 1987 |
| | Reclamation, land fill | Apr. 1986 | Oct. 1986 |
| EA | Pollution prevention projects | May 1985 | Dec. 1985 |
| MoT | Airports | Jan. 1986 | July 1986 |
| | Bullet trains | * | * |
| | Reclamation | Mar. 1986 | Sep. 1986 |
| Defence | Self-Defence Force airports | June 1987 | Dec. 1987 |
| MHW | Water treatment plants | Dec. 1985 | Dec. 1986 |
| | Water supply dams | Dec. 1985 | Dec. 1986 |
| NLA | Land development for Japan Regional Development Corporation | May 1987 | Sep. 1987 |
| MAFF | Reclamation | Oct. 1986 | Apr. 1987 |
| | Land development for agriculture | Feb. 1987 | Aug. 1987 |
| | Dams for agriculture | Nov. 1985 | May 1986 |
| MITI | Dams for industrial use | Mar. 1987 | Sep. 1987 |

Note: * indicates guidelines which existed before the Cabinet directive.
Source: Environment Agency, Coordination Bureau (22 June 1987).

**Table 6.3** EIAs reviewed by the EA

| | 1977 | 1978 | 1979 | 1980 | 1981 | 1982 | 1983 | 1984 | 1985 | 1986 | 1987 |
|---|---|---|---|---|---|---|---|---|---|---|---|
| *Based on Cabinet decision* | | | | | | | | | | | |
| Road | | | | | | | | | | 1 | 9 |
| Dam | | | | | | | | | | 1 | |
| Airport | | | | | | | | | | | 1 |
| Regional development | | | | | | | | | | | 2 |
| *Based on laws and ministry guidelines* | | | | | | | | | | | |
| Reclamation | 6 | 4 | 1 | 5 | 3 | 1 | 6 | 6 | 2 | 7 | 4 |
| Power plant | 28 | 45 | 22 | 49 | 33 | 30 | 20 | 11 | 23 | 24 | 6 |
| Port/harbour | 47 | 49 | 46 | 29 | 34 | 27 | 21 | 21 | 21 | 26 | 24 |
| Total | 89 | 98 | 69 | 83 | 70 | 58 | 47 | 38 | 46 | 59 | 46 |

Source: Sakurai (1988), p. 23.

Between 1977 and 1987, the EA reviewed almost 700 EIAs, as shown in Table 6.3. Of these, 50 per cent were based on the amendments to the Port and Harbour Law, 42 per cent on the MITI guidelines for power plant siting, 6 per cent on the Public Water Areas Reclamation Law, and 2 per cent on the Cabinet decision.

The EA does not plan to reintroduce an EIA bill to the Diet in the near future. Instead, it is promoting the preparation of REMPs by local authorities, as discussed in Chapter 4; EIA procedures would then 'harmonize' these plans with development plans.

### Local authority EIA systems

In the 1950s and 1960s many local authorities enacted pollution control laws, which in turn prompted the establishment of national pollution control laws. Similarly, local governments led the national government in the establishment of EIA systems. Table 6.4 and Figure 6.1 show trends in the institutionalization of local authority EIA systems.

The starting point for the establishment of local EIA procedures was the June 1972 Cabinet decision, which called for local governments to take environmental issues into consideration when implementing public works projects. Fukuoka prefecture set up EIA guidelines in 1973, followed by Tochigi prefecture in 1975 and Yamaguchi and Miyagi prefectures in 1976. These early guidelines were meant to be interim measures until the national government enacted legislation, and are correspondingly basic: they only require that the developer's proposal be reviewed by a prefectural environmental committee, and include no provisions for citizen participation or for the preparation of a final environmental impact statement (EIS).[19]

In response to increasing public pressure, the more progressive local authorities began to develop enforceable EIA ordinances. It was initially believed that local authorities with guidelines would later convert them to ordinances after they were found to be successful, but with the exception of Hokkaido this has not been the case.[20]

The first EIA ordinance was enacted by Kawasaki city in June 1976. Kawasaki is a heavily industrialized city of more than one million people at almost 1000 people/km$^2$, located between Tokyo (pop. 8 million) and Yokohama (pop. 2.8 million). Kawasaki's EIA procedures are strict. EIAs are required for a wide range of developments, even for those covering less than 1ha, and must include assessment of both environmental and social impacts. After a developer submits an EIA to the mayor, the EIA is made public, and citizens have a right to comment and demand a public hearing. The EIA is reviewed by an EIA committee. The mayor then comments publicly on the EIA, and may apply conditions to the development proposal.

**Table 6.4**  Local authority EIA systems

|  | Enacted | Operant |
|---|---|---|
| **Local authorities with EIA ordinances:** | | |
| Kawasaki city | Oct. 1976 | July 1977 |
| Hokkaido | July 1978 | Jan. 1979 |
| Tokyo | Oct. 1980 | Oct. 1981 |
| Kanagawa pref. | Oct. 1980 | July 1981 |
| **Local authorities with EIA guidelines:** | | |
| Fukuoka pref. | Apr. 1973 | Apr. 1973 |
| Tochigi pref. | Mar. 1975 | Mar. 1975 |
| Yamaguchi pref. | Jan. 1976 | Jan. 1976 |
| Miyagi pref. | May 1976 | May 1976 |
| Okinawa pref. | — | June 1977 |
| Kobe city | July 1978 | July 1978 |
| Okayama pref. | Dec. 1978 | Jan. 1979 |
| Nagoya city | Feb. 1979 | Apr. 1979 |
| Hyogo pref. | Mar. 1979 | Apr. 1979 |
| Mie pref. | Mar. 1979 | Apr. 1979 |
| Yokohama city | Jan. 1980 | Apr. 1980 |
| Kochi pref. | June 1980 | July 1980 |
| Nagasaki pref. | July 1980 | Aug. 1980 |
| Chiba pref. | Dec. 1980 | June 1981 |
| Saitama pref. | Feb. 1981 | June 1981 |
| Shiga pref. | Mar. 1981 | Mar. 1981 |
| Hiroshima pref. | Dec. 1982 | Apr. 1983 |
| Ibaraki pref. | Apr. 1983 | Oct. 1983 |
| Kagawa pref. | Sep. 1983 | Mar. 1984 |
| Nagano pref. | Jan. 1984 | Apr. 1984 |
| Osaka pref. | Feb. 1984 | Apr. 1984 |
| Aichi pref. | Mar. 1986 | Oct. 1986 |
| Kyoto pref. | May 1989 | Sep. 1989 |

Hokkaido enacted an EIA ordinance in July 1978. Hokkaido is a newly developing area with a low population density ($70/km^2$) and much open land. Its government was not so much concerned with pollution prevention, and infrastructure provision as with preserving the area's natural beauty while promoting sensible development. Consequently, its regulations require EIAs to be prepared for only a limited number of large-scale projects and to consider only a small range of impacts.

In October 1980 Kanagawa prefecture and Tokyo enacted EIA ordinances. Both areas are over-industrialized and have high population densities. The two EIA procedures resemble each other, and are noteworthy for the extent of their citizen participation procedures, with broad definitions of who can participate and mandatory public hearings. They also include post-EIA procedures for monitoring and feedback.[21]

**Figure 6.1** Status of local authority EIA systems

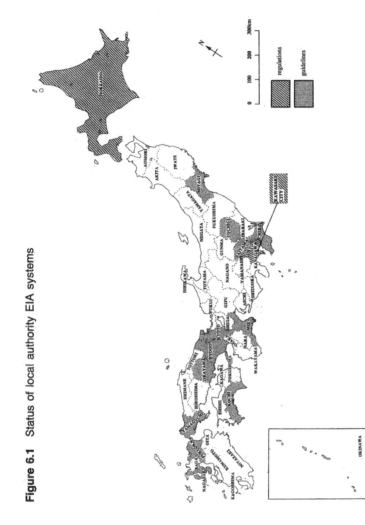

Source: Environment Agency, various publications.

No other local authority EIA ordinances have been enacted since then.

However, other local governments rapidly established EIA guidelines: Okinawa prefecture in 1977; Kobe city and Okayama prefecture in 1978; Nagoya city, and Hyogo and Mie prefectures in 1979; Yokohama and Nagasaki cities, and Kochi and Chiba prefectures in 1980; and Saitama and Shiga prefectures in early 1981. These guidelines generally incorporated citizen participation procedures similar to those of Kawasaki, and in some instances included post-assessment procedures.

Around this time the EA was encountering severe difficulties in its efforts to introduce EIA legislation, and began to survey the local authorities concerning EIA. These surveys showed that many local governments considered EIA to be a priority, and that local spending on EIA measures was increasing dramatically.[22] However, lacking national guidance, each local authority was developing its own unique set of EIA procedures without trying to unify or standardize them. This lack of uniformity caused problems, especially for projects which spanned more than one local authority. An EA survey of mid-1979 showed that many local authorities were requesting national leadership:

> Main problems involved in local governments' efforts to establish [EIA] procedures are the lack of widely accepted technical guidelines, . . . the type and scale of projects to be covered . . . and the manner and scope of public involvement in the assessment process. . . .
>
> As for local governments' requests to the national government, 24 local governments called for the establishment of a uniform legal procedure for [EIA], followed by 10 local governments which requested information on how to use various techniques, and 10 others which called for information regarding the progress of the study by the national government. . . .[23]

Five more prefectures set up EIA guidelines between 1981, when the national EIA bill was presented to the Diet, and the 1984 Cabinet decision: Hiroshima in 1982, Ibaraki and Kagawa in 1983, and Nagano and Osaka in 1984. These guidelines were strongly influenced by the proposed national EIA regulations, and adopted similar timescales and citizen participation procedures. However, they generally require EIAs to be conducted for a wider range of projects than earlier guidelines, including waste treatment plants and power stations.[24] Since the Cabinet decision, only Aichi and Kyoto prefectures have established EIA guidelines: these are based on those of the national government, but again cover a wider range of development projects.

Generally, however, the Cabinet decision has hindered further action by local authorities. The ministries' EIA guidelines 'request' that local governments 'harmonize' their procedures with the national guidelines, a logical request in light of the disunity of the many EIA systems. However, the national guidelines are weaker than most local guidelines, so local EIA systems are generally brought down to a lowest common denominator by this 'harmonization'. For instance, Kyoto city's attempts to establish EIA procedures for city planning have been consistently limited to those mandated by the MoC, which funds municipal development projects such as roads and subways.[25] Many local authorities which had planned to institute EIA systems have shelved these plans, including Wakayama, Niigata, and Toyama prefectures, and Kyoto city.

At present, twenty-seven of the fifty-seven prefectures and designated cities have established EIA procedures. In addition, Osaka city uses Osaka prefecture's guidelines, Hiroshima city uses Hiroshima prefecture's guidelines and Sapporo uses Hokkaido's regulations. Two smaller cities, Amagasaki and Yao, have also enacted guidelines (in November 1979 and August 1981 respectively). This leaves twenty-five prefectures and

**Table 6.5** EIAs reviewed by local authorities with EIA ordinances

|  | Operant | Dec. 1983 | Mid-1987 |
|---|---|---|---|
| Kawasaki | July 1977 | 30 | 52 |
| Hokkaido | Jan. 1979 | 33 | 45 |
| Tokyo | Oct. 1981 | 3 | 43 |
| Kanagawa | July 1981 | 5 | 25 |
| Total |  | 71 | 165 |

Sources: Environment Agency (1985), *Quality of the Environment in Japan 1984*, p. 91; Kajima Construction Co. (1987), pp. 27–36.

**Table 6.6** EIAs reviewed by local authorities, 1981–6

| Project type | 1981 | 1982 | 1983 | 1984 | 1985 | 1986 | Total |
|---|---|---|---|---|---|---|---|
| Road | 14 | 7 | 5 | 5 | 17 | 12 | 60 |
| Railway |  | 4 | 4 | 5 | 12 | 6 | 30 |
| Airport | 5 |  | 2 | 2 | 2 | 1 | 12 |
| Port/harbour |  | 2 |  | 1 | 1 |  | 4 |
| Reclamation | 9 | 20 | 21 | 23 | 32 | 9 | 114 |
| Dam | 1 | 3 | 1 | 1 | 2 |  | 8 |
| Residential | 9 | 11 | 12 | 15 | 18 | 15 | 80 |
| Waste treatment plant | 2 | 2 | 2 | 2 | 7 | 1 | 16 |
| Power station | 10 | 10 | 7 | 6 | 4 | 2 | 39 |
| Water supply works |  | 4 | 1 | 2 | 1 |  | 8 |
| Industrial | 1 | 6 | 2 | 6 | 2 | 1 | 18 |
| Soil excavation |  |  | 1 |  |  | 2 | 3 |
| Recreation | 1 | 3 | 2 | 1 | 13 | 5 | 25 |
| Other | 4 | 4 | 4 | 4 | 12 | 5 | 33 |
| Total | 56 | 76 | 64 | 72 | 123 | 59 | 450 |

Note: These data are only partial and should be taken as an indicative guide only.
Sources: Environment Agency (Mar. 1987), pp. 58–88; Kajima Construction Co. (1987), pp. 27–36.

two designated cities (Kyoto and Kita-Kyushu) which use the national guidelines only.

An EA survey showed that for the four local authorities with ordinances, a total of seventy-one EIAs were completed from the period of enactment to December 1983, and a further fourteen were still being prepared.[26] By mid-1987, these authorities had examined 165 EIAs, as shown in Table 6.5. Table 6.6 gives an indication of the EIAs reviewed by all local authorities between 1981 and 1986.

### Conclusions

With hindsight, one can see the difficulties that the EA faced in trying to implement EIA legislation, and conclude that the task it faced was almost impossible. Realistically, the best that the EA could hope to do was to develop some form of administrative coordination process which would facilitate the exchange of environmental information between government organizations. However, at the time the EA was young, idealistic, and flushed with its early successes: it did not realize the problems it faced in trying to impose a foreign legislation on a very strong, established, and factionalized administrative system. Starting from such an optimistic base allowed the EA to make a number of concessions:

— The consideration of alternatives was removed from the EIA process.
— Public participation measures were weakened.
— Judicial review was restricted.
— EIA was distanced from the possibility of lawsuits under pressure from the LDP.
— Power plants were removed from the list of projects subject to EIA under pressure from MITI and the electric power companies.
— City planning schemes were not subject to the Cabinet decision but are covered by MoC, which revised the City Planning Act to incorporate EIA requirements.

The development of Japan's EIA system illustrates a number of points about the administrative process:

— It highlights the difficulty of transferring environmental legislation from one country to another. The EA basically copied the NEPA approach to EIA, without trying to adapt it to Japan's very different political context and balance of power.
— It emphasises the difficulty of trying to reform administrative decision-making in Japan. The EA's efforts to institute environmental reforms through EIA, which alters traditional jurisdictional prerogatives, was opposed by other ministries, the LDP, the power industry and other commercial and industrial interests.

— It underlines the importance of consensus in the administrative process. The EA was unable to obtain support for its draft legislation because of its low standing in the ministerial hierarchy, and its lack of understanding of the political and jurisdictional difficulties involved. Nevertheless, the development of EIA guidelines by the MITI, MoC and MoT can be seen as an expression of a traditional willingness to cooperate and compromise.

— It demonstrates the complicated interrelationship of various environmental protection measures within the administrative process. For instance, guidance is often used to supplement regulations, and a draft legislation can be transformed into guidelines. Thus, although the EIA bill was defeated, many of its ideas and principles made their way into the national guidelines and local government EIA systems.

— It illustrates the influence which the judicial process, citizen protest and the activities of the opposition parties and local governments have on the administrative process, especially in the early phases of decision-making. Enactment of a mandatory national EIA system might have been possible in the climate of environmental consciousness and public activism of the late 1960s and early 1970s. However, as public priorities shifted away from environmental concerns and towards economic revitalization, the pressure exerted by these groups declined, and so did the central government's efforts to quickly enact EIA legislation.

## Notes

1. Huddle and Reich (1975), pp. 158–9.
2. Gresser *et al.* (1981).
3. Ibid., p. 277.
4. Ibid., p. 275. No mention is made of the environmental impact of operation activities as distinct from construction.
5. NEPA itself only included a vague reference to public involvement. However, this was sufficient to allow the development of NEPA by the courts and an active public into a major tool for environmental protection: Anderson *et al.* (1984), pp. 683–794.
6. Kurihara *et al.* (1982), pp. 289–90.
7. In 1987, this division had a staff of fifteen.
8. Gresser *et al.*, p. 276; Hase (1981b), p. 228.
9. E.g. violations of Kawasaki's ordinance are punishable with a ¥50,000 ($400) fine; Gresser *et al.* (1981), p. 278.
10. Isobe *et al.* (1979), pp. 15–30.
11. Gresser *et al.* (1981), p. 276; Kihara (1981), p. 505.
12. At a meeting with the LDP committees, the EA's director proposed to abolish the compensation scheme for pollution victims in return for the establishment of an EIA system. The LDP reportedly asked for the abolishment of *both* systems. . . .
13. Isobe *et al.* (1979), p. 19.

14. Suzuki (1985), p. 54.
15. Ibid., pp. 54–61.
16. Environment Agency (Nov. 1984).
17. Committee for Fostering the Implementation of Environmental Impact Statements (Nov. 1984).
18. Environment Agency (Feb. 1985).
19. Hase (1981b), p. 232.
20. Ibid.
21. Kajima Construction Co. (1987), pp. 102–3.
22. For example, between 1977 and 1978, spending on the implementation of EIA by prefectures and major cities increased from ¥420 million to ¥690 million, and spending on 'improvements to the system' increased from ¥26 million to ¥151 million: Environment Agency (Oct. 1978) p. 243.
23. Environment Agency (Nov. 1979), p. 330.
24. Kajima Construction Co. (1987), pp. 24–6, 102–3.
25. In December 1977 Kyoto city's Pollution Control Division released a report on items that should be considered when developing an EIA system for Kyoto. In 1980 the city asked the Pollution Affairs Council, an advisory body, to consider the possibility of introducing an EIA system. The council recommended an outline system in January 1981. The Pollution Control Division wanted to institute guidelines based on this outline, and presented the proposal to the city's Sanitation, City Planning and Construction departments, and to the MoC, whose approval it required. The MoC vetoed the proposal on the grounds that it did not comply with national guidelines.
26. Environment Agency (1987), p. 91; Kajima Construction Co. (1987), pp. 27–36.

# Chapter seven

# Procedures for environmental impact assessment

## Introduction

The previous chapter discussed how the EA's failure to produce effective national EIA legislation led to the development of a number of guidelines and regulations by ministries and local authorities. This chapter examines Japan's EIA procedures. First the procedures specified in the 1984 Cabinet decision and related ministerial guidelines are discussed. Then EIA procedures at the local authority level are reviewed. Finally, additional EIA procedures prescribed in regulations and ministry guidelines are discussed.

## National EIA procedures

The Cabinet decision of August 1984 set out a scheme for the assessment of certain types of projects with which the state is concerned. It requested that national administrative agencies ensure that the projects they undertake are preceded by an EIA. Local authorities and industry are subject to the scheme when they require licences, approval or guidance from a national ministry. However, although the Cabinet decision purported to allow the local authorities freedom to establish their own EIA procedures, it required that these procedures 'pay due attention to conformity' with the national scheme.[1] A translation of the Cabinet decision is given in Appendix C.

### Summary of procedures

Figure 7.1 summarizes the national EIA procedures. The procedures are:

Prior to construction, if a development is subject to the EIA guidelines, the developer should investigate conditions at the site in accordance with the technical guidelines provided by the responsible ministry. The developer should assess the project's environmental impact and consider possible pollution control measures.

**Figure 7.1**   Summary of national EIA procedures

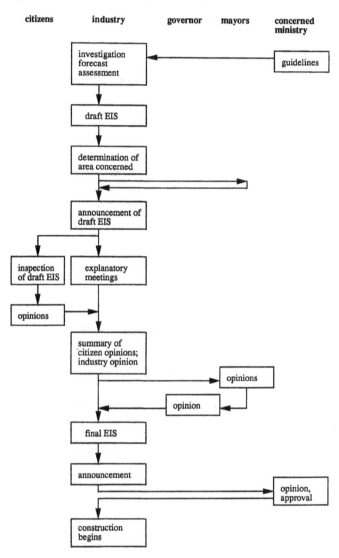

— The developer should then compile a draft EIS, determine the area over which the project will have an impact, and present the EIS to the governor and mayors of that area. The draft EIS should be made publicly available for one month. During that time the developer should, if possible, hold explanatory meetings for residents of the affected area.[2]

— For one calendar month plus two weeks after the release of the draft EIS, residents of the affected area can send written comments to the developer. The developer should summarize these and present them to the governor and mayors.

— Within three months of receiving the residents' opinions, the governor, in consultation with the mayors, should comment on the EIS from the perspective of pollution control and conservation of the natural environment.

— After receiving the governor's opinion, the developer should prepare a final EIS, including any revisions of information from the draft EIS, a summary of residents' opinions, the governor's opinion, and measures taken to respond to these opinions. The developer should make the final EIS available to the public for one month.

— If licence or permit approvals are involved, the developer should present the approved EIS to the responsible agency, which must

**Table 7.1** Approach to EIA procedures

| | |
|---|---|
| *Investigation* | |
| Items to be surveyed | Ministry guidelines |
| Survey techniques | Ministry guidelines |
| Survey duration/frequency | Ministry guidelines |
| Survey area | Area where the environment will be changed by the project |
| *Forecast* | |
| Items to be predicted | As deemed necessary from analysis of survey data |
| Prediction techniques | Ministry guidelines |
| Prediction of pollution control | Simulation models, experiments, reference to previous cases |
| Prediction of nature conservation | |
| Prediction timescale | Ministry guidelines |
| Prediction area | Within area of survey |
| *Assessment* | |
| Influence of other projects in survey area | Results of surveys/predictions, proposed pollution control measures |
| Assessment techniques | Ministry guidelines |
| Protection of human health | Environmental quality standards |
| Conservation with respect to importance of natural environment | Best available scientific knowledge |

**Table 7.2**   Projects subject to national EIA procedures

| Implement-ation scheme | Projects | Project Undertakers | | | Competent ministry |
|---|---|---|---|---|---|
| | | 1. State, Public Corporations (except 2 and 3) | 2. Those involved in permission, approval or reporting | 3. Those involved in subsidization (except 2) | |
| (1) Roads | Construction or reconstruction of national expressway | MoC | | | |
| | Construction of national road or by-passes with ≥4 lanes, widening of national road for addition of ≥4 lanes (except if <10km) | MoC | Japan Highway Public Corpora-tion, Honshu–Shikoku Bridge Authority, local road develop-ment public corpora-tions | Prefectural governors, mayors of designated cities | MoC |
| | Construction or reconstruction of Metropolitan and Hanshin expressways, and expressways in designated cities with ≥4 lanes | | Metro-politan Express-way Public Corpora-tion, Hanshin Express-way Public Corpora-tion, local road develop-ment public corp. | | |
| (2) Dams, rivers | Construction of dams with a water surface area of >200ha, on class 1 rivers | MoC | Prefectural governors | | MoC |
| | | | WRDC, those other than local public bodies | Local public bodies | MITI |
| | | | WRDC, municipal-ities | | MHW |

| Implement-ation scheme | Projects | Project Undertakers | | | Competent ministry |
| | | 1. State, Public Corporations (except 2 and 3) | 2. Those involved in permission, approval or reporting | 3. Those involved in subsidization (except 2) | |
| --- | --- | --- | --- | --- | --- |
| | | MAFF | WRDC | Prefectures, municipalities | MAFF |
| | Development of lakes and construction of drainage canals involving a modified land area of ≥100ha | MoC | Prefectural governors | | MoC |
| (3) Railways | Construction and large-scale improvement of Shinkansan railways | | JNR JRCPC | | MoT |
| (4) Airports | Construction of airports with runway length of ≥2,500m, construction of new runways of ≥2,500m, extension of runways by ≥500m (excluding cases in which total runway length after extension is <2,500m) | MoT Defence Agency | New Tokyo Inter-national Airport Authority, local public bodies | | MoT Defence Agency |
| (5) Reclama-tions and dumping | Final disposal sites for domestic and industrial wastes, covering an area of ≥30ha at completion of construction or modification | | Municipal-ities, project under-takers | | MHW |
| | Reclamation or drainage of public water areas covering ≥50ha | | Project under-takers | | MoC MoT |

*contd.*

**Table 7.2**   continued

| Implement-ation scheme | Projects | Project Undertakers | | | Competent ministry |
| | | 1. State, Public Corporations (except 2 and 3) | 2. Those involved in permission, approval or reporting | 3. Those involved in subsidization (except 2) | |
| --- | --- | --- | --- | --- | --- |
| | Reclamation or drainage covering ≥50ha undertaken as land improvement projects | MAFF | | Prefectures | MAFF |
| (6) Land readjust-ment | Land readjustment projects covering ≥100ha | MoC | Prefec-tures, municipal-ities, HUDC, JRDC, LHDPCs | Land readjustment association, etc. | MoC |
| (7) New residential area develop-ment | New residential area development projects covering ≥100ha | | Prefec-tures, municipal-ities, HUDC, JRDC, LHDPCs | | MoC |
| (8) Industrial estate construc-tion | Construction projects of industrial estate covering ≥100ha prescribed by the Act concerning the Development of Suburban Development and Redevelopment Areas in the National Capital Region and the Act concerning the Development of Suburban Development and Redevelopment Areas in the Kinki Region | | Prefec-tures, municipal-ities, HUDC, JRDC | | MoC |

| | | Project Undertakers | | | |
|---|---|---|---|---|---|
| Implement-ation scheme | Projects | 1. State, Public Corporations (except 2 and 3) | 2. Those involved in permission, approval or reporting | 3. Those involved in subsidization (except 2) | Competent ministry |
| (9) New urban infra-structure develop-ment | New urban infrastructure development projects covering ≥100ha | | Prefec-tures, municipal-ities, HUDC, JRDC | | MoC |
| (10) Distribution business develop-ment | Construction of distribution business centres covering ≥100ha under the Law for the Development of Distribution Business Centres | | Prefec-tures, municipal-ities, HUDC, JRDC | | MoC |
| (11) Land creation by public corpora-tions | Creation of housing sites covering ≥100ha under Art.29, para 1.2, 1.5, or 1.15a of the HUDC Law | HUDC | | | MoC |
| | Creation of land covering >100ha under Art.19, para.1.1 of the JRDC Law | | JRDC | | NLA, MoC |
| | Land creation projects covering ≥100ha under Art.19, para. 1.3 of the JRDC Law | JRDC | JRDC | | MITI<br>NLA, MITI |
| | Land creation projects covering ≥100ha under Art.19, para. 1.4 of the JRDC Law | | JRDC | | NLA, MITI |
| | Land creation projects covering ≥100ha under Art.18, para 1 or 3 of the Pollution Control Services Corporation Law | | Pollution Control Services Corporation | | EA |

*contd.*

**Table 7.2**   continued

| | | Project Undertakers | | | |
|---|---|---|---|---|---|
| Implement- ation scheme | Projects | 1. State, Public Corporations (except 2 and 3) | 2. Those involved in permission, approval or reporting | 3. Those involved in subsidization (except 2) | Competent ministry |
| | Land creation projects for agricultural use covering ≥500ha within areas of projects based on Art.19, para. 1.1a of the Agricultural Land Development Public Corporation Law | | Agricultural Land Develop- ment Public Corporation | | MAFF |

*Key*: HUDC – Housing and Urban Development Corporation; JNR – Japanese National Railways; JRCPC – Japan Railway Construction Public Corporation; JRDC – Japan Regional Development Corporation; LHDCPs – Local Housing Development Public Corporations; LRDPCs – Local Road Development Public Corporations; WRDC – Water Resources Development Corporation.

ensure that the EIS properly considers matters related to pollution control and nature conservation. The ministry may solicit the views of the EA if special environmental consideration is needed.

— When local authorities are the developer and state subsidies are involved, the national government must consider the expenses needed to implement the required EIA procedures.[3]

Table 7.1 shows the survey, prediction and evaluation techniques prescribed for the assessments. Analysis techniques are specified in the ministry guidelines: environmental quality standards and the best scientific knowledge available should be used.

*Range of projects subject to EIA*

Table 7.2 lists the types of projects which are considered to have a marked effect on the environment and which involve national govern- ment participation by, e.g., licence approval or subsidies. An additional category, other projects similar to those in Table 7.2, can be subject to assessment if designated by the responsible ministry and the EA's director: thus the national government can claim that virtually all projects are subject to EIA if the administration so desires.

EIAs must be prepared for the following project stages:

for all projects except landfill or reclamation:
  (a) construction/execution of the project,
  (b) the land/structures completed by (a), and
  (c) activities conducted on/in (b).
for landfill or reclamation projects:
  (a) construction/execution of the project, and
  (b) the land/structures (including wharves, banks and similar struc-
      tures) completed by (a).[4]

*Scope of the EIA*

The focus of the national EIA guidelines is on pollution control and
conservation of the natural environment. Table 7.3 lists the factors
included under these categories.

*Public participation procedures*

The Cabinet decision includes only a restricted definition of who can
participate in EIA procedures, and of how they can participate. The
developer determines the area over which the proposed project will have
an impact, and the residents and corporations located in that area can
participate. Concerned residents can send written comments on the draft
EIS to the governor, and the developer must explain the draft EIS at
public meetings.

However, aside from this few forums exist for public participation in
environmental decision-making. Petitions or meetings with government
or corporate representatives are often viewed as the complaints of a
radical minority and ignored. There are no procedures for public hearings
(although hearings are sometimes held for large projects), for citizen
participation in advisory committees, or for citizens to comment on the

**Table 7.3**  Impacts considered in the national EIA procedures

| Pollution prevention | Nature conservation |
|---|---|
| Air pollution | Topography |
| Water pollution | Geology |
| Soil pollution | Plants |
| Noise | Animals |
| Vibration | Scenery |
| Ground subsidence | Outdoor recreation |
| Offensive odour | |

final EIS. International pressure is proving to be effective in influencing decision-making, since Japan is currently concerned about 'internationalization'; an example of this is the New Ishigaki Airport, discussed in Chapter 12. However, scientific or politically powerful groups are more effective at this than are citizens or environmentalists.

*Administrative review*

The Cabinet decision sets up procedures for three types of public review which are expected to ensure that industry exercises 'self-control': the governor's opinions, which are incorporated into the EIS and scrutinized by the public; the concerned ministry's check of the final EIS; and the views of the EA, via the concerned ministry. These procedures will be discussed further in Chapter 8.

### Local authority EIA procedures

Lacking national guidance, the local authorities devised EIA systems which best suited their requirements, while trying to anticipate and comply with any future national EIA legislation. The result is a heterogeneous jumble of decisions, regulations and guidelines which range from virtually complete government discretion (and, in practice, industrial *laissez-faire*) in Okinawa, to Kawasaki's and Tokyo's extremely detailed EIA procedures.

The local procedures are generally more demanding than the national ones, and give the local governor many of the responsibilities that the national procedures give to the developer. Local procedures also tend to include more opportunity for public involvement and are more detailed than the national procedures, both in the types of projects that require appraisal and in the range of environmental factors which must be considered.

Local authority EIA systems can be sub-divided into three groups:

— *Interim* (before 1976). This group consists of early guidelines which were originally envisaged as interim measures until the enactment of national legislation. These guidelines have no procedures for public involvement and only minimal administrative review procedures. This group includes Fukuoka, Tochigi, Yamaguchi, Miyagi, Okinawa, Kochi and Nagasaki; the last two authorities enacted procedures in 1980, but the procedures' relative simplicity merits their inclusion in this group
— *Extensive* (1976–81). This group marks the high point of local autonomy before the national EIA bill was introduced to the Diet in 1981. It includes the four authorities which instituted EIA ordinances

– Kawasaki, Hokkaido, Tokyo and Kanagawa – and those whose guidelines include extensive citizen participation procedures, pre- and post-EIA procedures, and enforcement measures: Kobe, Okayama, Nagoya, Hyogo, Mie, Yokohama, Chiba, Saitama and Shiga.

— *Uniform* (after 1981). This group consists of seven authorities which established guidelines after the EIA bill was introduced to the Diet: Hiroshima, Ibaraki, Kagawa, Nagano, Osaka, Aichi and Kyoto. Their guidelines, influenced by the national bill and presumably striving for unification with any future national legislation, are much more uniform than those of the previous two groups. They give more power to the developer, incorporate less public involvement, and are all guidelines rather than ordinances.

### Summary of procedures

The local authority procedures for EIA are summarized in Table 7.4. The procedures can be divided into eight main phases:

— pre-EIA procedures,
— preparation of a draft EIS, explanatory meeting, and citizen comments,
— public hearing,
— response to citizen comments,
— second public hearing,
— opinion of the review committee,
— preparation of the final EIS, and
— post-EIA procedures.

Local authority procedures vary greatly, but can be summarized as follows. In the pre-EIA phase, the developer notifies the local authority of its intention to undertake an EIA for a proposed project. A draft EIS is prepared and released for public inspection for thirty days. During this time the residents of the concerned area (definition varies by authority) can comment on the EIS. The governor or developer then compiles a report summarizing these comments, and this is released to all parties for forty-five days. A public hearing may be held if called for by the governor. Next the developer must respond to the residents' comments. The second report is made public for fifteen to twenty days, and in some cases local officials can receive written comments for twenty to thirty days. A second public hearing may be held. An administrative review committee then reports to the governor. The developer prepares a final EIS which is subject to inspection for one week to a month. In some instances post-EIA studies are required.[5]

**Table 7.4** Local authority EIA procedures (national guideline included for comparison)

| Date established<br>Status | Fukuoka<br>Apr. 1973<br>guideline | Tochigi<br>Mar. 1975<br>policy | Yamaguchi<br>Jan. 1976<br>notification | Miyagi<br>May 1976<br>guideline | Kawasaki<br>Oct. 1976<br>ordinance |
|---|---|---|---|---|---|
| *Pre-EIA formalities* | | | | | |
| Report on execution of project | D | | | | |
| Report on EIA commencement | | | | | |
| Notification of EIA execution plan | | | | | |
| Public inspection of EIA implementation | | | | | |
| *Draft EIS* | | | | | |
| Preparation of draft EIS | | D | D | D | D |
| Public notification and inspection | | | | | M 30d |
| Release of established plan | | | | | D |
| Explanatory meeting | | | | | D |
| Scope of citizen opinion | | | | | B |
| Citizen opinion sent to | | | | | M 45d |
| *Public hearing* | | | | | |
| Request for hearing | | | | | |
| Hold public hearing | | | | | |
| *Response to citizen opinion* | | | | | |
| Prepare report on citizen opinion | | | | | D |
| Public inspection of report | | | | | M 15d |
| Explanatory meeting | | | | | |
| Officials receive written comments | | | | | |
| *Public hearing* | | | | | |
| Request for hearing | | | | | M |
| Hold public hearing | | | | | Mx |
| *Opinions of deliberation committee* | | | | | |
| Composition of committee | | | | | 20C |
| Deliberation questions | | | | | M |
| Opinions of concerned mayors | | | | | |
| Report of deliberation committee's opinion | G | G | G | G | M |
| Public inspection of report | | | | | M |
| *Final EIS* | | | | | |
| Preparation of final EIS | | | | | |
| Public inspection of final EIS | | | | | |
| *Post-EIA procedures* | | | | | |
| Release of monitoring plan | | | | | |
| Notice of monitoring results | | | | | Dx |
| Public inspection of results | | | | | |
| Report commencement of construction | | | | | |
| Report completion | | | | | |
| *Other* | | | | | |
| Coordination with national government | | | | | — |
| Coordination with city plans, etc. | | | | | — |

| Okinawa June 1977 policy | Kobe July 1978 guideline | Hokkaido July 1978 ordinance | Okayama Dec. 1978 guideline | Nagoya Feb. 1979 guideline | Hyogo Mar. 1979 guideline | Mie Mar. 1979 guideline | Yokohama Jan. 1980 policy |
|---|---|---|---|---|---|---|---|
| | | | D | | D | | |
| | | | | | | D | |
| | D | | | D | | | |
| | G 2w | | | G 15d | | | |
| | D | D | D | D | D | D | D |
| | M 1m | G 30d | G | M 30d | D 30d | D 1m | M 30d |
| | D | | | D | | | D |
| | D | Gx | Dx | D | D | D | D |
| | B | A | B | B | B | A | B |
| | M 45d | G 45d | D | M 45d | G/D 30d | D 1m+2w | M 45d |
| | M | | | | | | |
| | Mx | | | | Gx | | |
| | D | | | D | D | | D |
| | | | | M 15d | | | M 15d |
| | | | | | | | M 20d |
| | | | | M | | | |
| | | Gx | | Mx | | | |
| | 23S | 20S | 15S | 20S | 15S | 15S | 15S |
| | M | G | | M | G | Gx | M |
| | M | G | G | M | G | G | M |
| | M 2w | G | | M 15d | | | M |
| | D | D | D | D | D | D | D |
| | M 2w | G | | M 7d | G 15d | D 1w | M |
| | | | Dx | D | | | |
| | Dx | | Dx | D | | | Dx |
| | M 2w | | | M 7d | Dx | Gx/Dx | D |
| | | | | D | | | D |
| | | | | D | | | |
| — | — | — | — | — | — | — | — |
| | | | | | — | — | — |

**Table 7.4**   continued

| Date established<br>Status | Kochi<br>July 1980<br>guideline | Nagasaki<br>July 1980<br>guideline | Tokyo<br>Oct. 1980<br>ordinance | Kanagawa<br>Oct. 1980<br>ordinance | Chiba<br>Dec. 1980<br>guideline |
|---|---|---|---|---|---|
| *Pre-EIA formalities* | | | | | |
| Report on execution of project | | | | | D |
| Report on EIA commencement | D | | | | |
| Notification of EIA execution plan | | | | | |
| Public inspection of EIA implementation | | | | | G |
| *Draft EIS* | | | | | |
| Preparation of draft EIS | D | D | D | D | D |
| Public notification and inspection | | Dx | G 30d | G 30d | G 30d |
| Release of established plan | | | D | D | |
| Explanatory meeting | | | D | D | D |
| Scope of citizen opinion | | A | Tokyo Res. | B | A |
| Citizen opinion sent to | | Dx | G 45d | G 45d | G 45d |
| *Public hearing* | | | | | |
| Request for hearing | | | | | |
| Hold public hearing | | | G | | |
| *Response to citizen opinion* | | | | | |
| Prepare report on citizen opinion | | | D | D | D |
| Public inspection of report | | | G 20d | G 20d | |
| Explanatory meeting | | | D | | |
| Officials receive written comments | | | G 30d | G 30d | |
| *Public hearing* | | | | | |
| Request for hearing | | | | | |
| Hold public hearing | | | | G | Gx |
| *Opinions of deliberation committee* | | | | | |
| Composition of committee | | | 20S | 20S | 15S |
| Deliberation questions | | | G | G | G |
| Opinions of concerned mayors | | | | | |
| Report of deliberation committee's<br>   opinion | G | G | G | G | G |
| Public inspection of report | | | | | |
| *Final EIS* | | | | | |
| Preparation of final EIS | D | | D | D | D |
| Public inspection of final EIS | | | G 15d | G 15d | G 15d |
| *Post-EIA procedures* | | | | | |
| Release of monitoring plan | | | D | | |
| Notice of monitoring results | | | D | Gx/Dx | Dx |
| Public inspection of results | | | G | | |
| Report commencement of construction | | | D | D | |
| Report completion | | | D | D | |
| *Other* | | | | | |
| Coordination with national government | | — | — | | — |
| Coordination with city plans, etc. | | — | — | — | — |

| Saitama Feb. 1981 guideline | Shiga Mar. 1981 guideline | Hiroshima Dec. 1982 guideline | Ibaraki Apr. 1983 guideline | Kagawa Sep. 1983 guideline | Nagano Jan. 1984 guideline | Osaka Feb. 1984 guideline | National Aug. 1984 guideline |
|---|---|---|---|---|---|---|---|
|  | D |  |  |  | D |  |  |
|  |  |  |  | D |  | D |  |
| D | D | D | D | D | D | D | D |
| G 30d | G 30d | G 30d | G 30d | G 1m | G 1m | G 1m | G/D 1m |
|  | D |  |  |  |  |  |  |
| D | D | D | D | D | D | D | D/Gx |
| A | B | A | A | A | A | B | A |
| G 45d | G 45d | D 45d | D 45d | D 1m+2w | D 1m+2w | G/D 1m+2w | D 1m+2w |
| D |  |  |  |  |  | Dx | D |
|  |  |  |  |  |  | G |  |
| G | Gx | Gx | Gx | Gx | Gx | Gx |  |
| 15S | 10S |  | 15S | 10S | 10S | 20S |  |
| G | G |  | G | Gx | G | Gx |  |
|  |  |  |  |  |  |  | G |
| G | G | G | G | G | G | G |  |
|  | G |  |  |  |  |  |  |
| D | D | D | D | D | D | D | D |
| G 15d | G 30d | G 30d | G 30d | G 1m | G 1m | G 1m | D/G 1m |
| Dx | Dx | Gx/Dx |  | Dx | Gx/Dx |  |  |
| D | D |  |  |  | D |  |  |
| D |  |  |  |  | D |  |  |
| — | — | — | — | — | — | — | — |
| — | — | — | — | — |  | — |  |

**Table 7.4** continued

| Date established<br>Status | Aichi<br>Mar. 1986<br>guideline | Kyoto<br>May 1989<br>guideline |
|---|---|---|
| *Pre-EIA formalities* | | |
| Report on execution of project | | |
| Report on EIA commencement | | |
| Notification of EIA execution plan | | |
| Public inspection of EIA implementation | | |
| *Draft EIS* | | |
| Preparation of draft EIS | D | D |
| Public notification and inspection | G 1m | G 1m |
| Release of established plan | | |
| Explanatory meeting | D | D |
| Scope of citizen opinion | A | A |
| Citizen opinion sent to | D 1m+2w | D 1m+2w |
| *Public hearing* | | |
| Request for hearing | | |
| Hold public hearing | | |
| *Response to citizen opinion* | | |
| Prepare report on citizen opinion | | D |
| Public inspection of report | | |
| Explanatory meeting | | |
| Officials receive written comments | | |
| *Public hearing* | | |
| Request for hearing | | |
| Hold public hearing | | Gx |
| *Opinions of deliberation committee* | | |
| Composition of committee | 20S | na |
| Deliberation questions | Gx | G |
| Opinions of concerned mayors | | |
| Report of deliberation committee's<br>  opinion | G | G |
| Public inspection of report | | |
| *Final EIS* | | |
| Preparation of final EIS | D | D |
| Public inspection of final EIS | G 1m | G 1m |
| *Post-EIA procedures* | | |
| Release of monitoring plan | | |
| Notice of monitoring results | | |
| Public inspection of results | | |
| Report commencement of construction | | |
| Report completion | | |
| *Other* | | |
| Coordination with national government | — | — |
| Coordination with city plans, etc. | — | — |

Key: A – concerned area's citizens; B – anyone with an opinion; C – specialists and citizens; D – developer; Dx – developer, if necessary; G – governor; Gx – governor, if necessary; M – mayor; Mx – Mayor, if necessary; S – specialists only; d – day; w – week; m – month; * – coordination provisions; na – not available.

Sources: Kajima Construction Co. (1987) pp. 102–3; Therivel (1988), pp. 64–6.

*Range of projects subject to EIA*

Table 7.5 summarizes the developments subject to EIA under local ordinances and guidelines. The range of developments subject to EIA varies widely according to the local authority circumstances and the characteristics of the area. In Hokkaido, for instance, only eight types of large-scale projects are subject to assessment, whereas in overcrowded Tokyo EIAs are required for twenty-six types of development.

The interim group's EIA procedures generally apply to only a limited number of projects, but the extensive and uniform groups' generally apply to a wider range of developments than do the national guidelines. Local procedures also often apply to smaller projects than do the national guidelines. The national procedures apply only to projects with an area of ≥100ha, whereas local procedures can apply to projects as small as 1ha or less.

*Scope of the EIA*

Table 7.6 lists the types of environmental impacts which must be assessed according to the extensive and uniform groups. In general, the local systems require more environmental factors to be assessed than do the national guidelines. The more urbanized authorities include provisions for factors such as sunshine impediment, interference with television reception, and the maintenance of mixed scenery and green areas. Most local authorities also require that socio-cultural factors be considered, although none require a comprehensive assessment of the socio-economic costs and benefits of a project. Several authorities require the considera-tion of natural accidents such as landslides and floods, and man-made accidents such as traffic problems, fires and explosions.

*Public participation procedures*

Although the interim group has no procedures for citizen participation, the extensive and uniform groups do. Generally, the extensive group allows anyone who is interested to participate, whereas the uniform group limits participation to residents of the affected area. The size of the affected area is determined differently from authority to authority, but the developer and local government usually cooperate in the decision. Most extensive group procedures stipulate that the governor should receive the citizen opinions, but the uniform groups resemble the national guidelines in requiring developers to receive them.

During the time when the draft EIS is open to the public, the developer has to hold explanatory meetings. Tokyo's regulations require that the developer must hold two rounds of meetings, the latter to explain

**Table 7.5**   Projects subject to local and national EIA procedures ordinances

| Date enacted: | *National*<br>*Aug. 1984* | *Tokyo*<br>*Oct. 1980* | *Kanagawa*<br>*Oct. 1980* |
|---|---|---|---|
| **Roads**<br>national expressways | All (excluding small scale<br>such as interchanges) | All new development<br>over 1km | All new development |
| others | 4 lane over 10km | 4 lane over 1km | "      " |
| **Dams**<br>Lakes | 200ha | Hgt.15m, 100ha | Height over 15m |
| Rivers | 100ha | | |
| Railways | Bullet train | All development<br>(incl. monorail) | 1km |
| Airports | 2,500m runway | All new airports | 1ha |
| Land reclamation | 30ha waste disposal | 15ha | 15ha (1ha) |
| Dumping | 50ha public water areas<br>50ha land improvement | | |
| Soil extraction | | 10ha | 10ha(1ha) |
| New residential | 100ha | | |
| Industrial estate | 100ha | all | 10ha (1ha) |
| New urban infrastructure | 100ha | all | |
| Distribution business centres | 100ha | all | |
| HUDC housing development | 100ha | | |
| JRDC land preparation | 100ha | | |
| EPSC land creation | 100ha | | |
| Land creation for<br>  agricultural use | 500ha | | |
| Housing redevelopment | | 40ha (1000 houses) | 20ha (1ha) |
| Factories | | 9,000m² (3000m²) | 3ha (1ha) |
| Ports and harbours | | 240m | n.a. |
| **Waste treatment plants**<br>garbage | | 200 tons per day | 3ha (1ha) |
| sludge | | 100kl per day | "      " |
| waste disposal (fill) | 30ha | 1ha (50,000m³) | "      " |
| Water supply | | | 10ha (1ha) |
| Sewage | | 5ha | |
| Recreation projects | | 40ha | 20ha (1ha) |
| **Power plants**<br>hydro | Covered by separate | 30,000kW | 30,000kW |
| thermal | procedures administered | 150,000kW | 150,000kW |
| nuclear | MITI | all | all |
| geotherm | | | 10,000kW |
| Other | | Gas plants: 1,500,000m³<br>per day<br>Oil pipeline: 15km<br>Oil plant: 30,000kl<br>Car park: 1000 cars<br>High buildings: 100m | Railyard: 10ha (1ha)<br>Research centre:<br>3ha (1ha)<br>Cemetery: 20ha (1ha) |

| Date enacted: | Hokkaido (Sapporo) July 1978 | Kawasaki Oct. 1976 | Fukuoka Apr. 1974 | Tochigi Mar. 1975 |
|---|---|---|---|---|
| **Roads** | | | | |
| national expressways | All new development over 5km, W.5.5m | All new development with 4 lanes | | |
| others | 4 lane and 2km | "       " | | |
| **Dams** | | | | |
| Lakes | 200(30)ha | | | |
| Rivers | | | | |
| Railways | Bullet train | Not specified | | |
| Airports | 2km runway 0.5km extension | | | |
| Land reclamation | | 1ha | | |
| Dumping | | | 3ha | |
| Soil extraction | | | 3ha | |
| New residential | 100ha | 1ha:500 people | 3ha | 50ha |
| Industrial estate | 100ha | site prep: 9,000m$^2$ constr. 3,000m$^2$ | | 20ha |
| New urban infrastructure Distribution business centres HUDC housing development JRDC land preparation EPSC land creation | | | | |
| Land creation for agricultural use | | | | |
| Housing redevelopment | | | | |
| Factories | | 9,000m$^2$ (3000m$^2$) | | |
| Ports and harbours | | breakwater | | |
| **Waste treatment plants** | | | | |
| garbage | | 9,000m$^2$ (3000m$^2$) | | |
| sludge | | "       " | | |
| waste disposal (fill) | | "       " | | |
| Water supply | | Not specified | | |
| Sewage | | | | |
| Recreation projects | 300ha | | Golf course 3ha | 50ha |
| **Power plants** | | | | |
| hydro | 30,000kW | | | |
| thermal | 15,000kW | | | |
| nuclear | all | | | |
| geotherm | 10,000kW | | | |
| Other | | Gas plants 0.9ha Water purification plants | Mineral working: 3ha | Other |

**Table 7.5**   continued

| Date enacted: | Yamaguchi Jan. 1976 | Miyagi May 1976 | Okinawa June 1977 | Kochi June 1980 |
|---|---|---|---|---|
| Roads national expressways | | | | |
| others | | 4 lane over 2km | | |
| Dams Lakes Rivers | | | | |
| Railways | | | | Bullet train |
| Airports | 50(20)ha | | | Not specified |
| Land reclamation Dumping | 15ha | 20ha | | Not specified |
| Soil extraction | | | | |
| New residential | 50(20)ha | 50(20)ha | | MoC jurisdiction 300ha |
| Industrial estate | 50(20)ha | 50(20)ha | | 10ha |
| New urban infrastructure Distribution business centres HUDC housing development JRDC land preparation EPSC land creation Land creation for agricultural use | | | | |
| Housing redevelopment Factories | | | | |
| Ports and harbours | | Cost >¥5 billion | | Not specified |
| Waste treatment plants garbage sludge waste disposal (fill) | | 10ha | | |
| Water supply Sewage | | 20ha | | |
| Recreation projects | 50(20)ha | 20ha | | |
| Power plants hydro thermal nuclear geotherm | | | | |
| Other | Livestock: 20ha River and water resource development: 50(20)ha | Livestock River and water resource development Other | | Plans and alterations under City Planning Law: Other |

| Date enacted: | Nagasaki July 1980 | Kobe July 1978 | Okayama Dec. 1978 |
|---|---|---|---|
| **Roads** national expressways | | | |
| others | 4 lane/10km (2 lane) | 4 lane/3km (2 lanes) | 4 lanes (2 lanes) |
| **Dams** Lakes Rivers | 30ha | | 50ha |
| Railways | Not specified | Not specified | Not specified |
| Airports | All new 500m extension | All new | All new |
| **Land reclamation** Dumping | Drainage: 5ha Land fill: 15ha | PWAR Law | 10ha |
| Soil extraction | | | |
| New residential | 50ha | 20ha | 20ha |
| Industrial estate | 30ha | 10ha | 10ha |
| New urban infrastructure Distribution business centres HUDC housing development JRDC land preparation EPSC land creation Land creation for agricultural use | 30ha | 20ha | |
| Housing redevelopment Factories | | 20ha | |
| Ports and harbours | | 1km breakwater | |
| **Waste treatment plants** garbage sludge waste disposal (fill) | 100T/d 300kl/d 1ha | 200T/d 15ha | 10ha |
| Water supply Sewage | WW: 5,000m³/d Sewage: 50,000m³/d | WW 10,000m³/d Not specified | WW 10,000m³/d Not specified |
| Recreation projects | | 20ha | 10ha |
| **Power plants** hydro thermal nuclear geotherm | Not specified | Not specified | Not specified |
| Other | Energy saving basis Gas: 100,000Nm³/hr | Gas: 40,000Nm³/hr Electricity/gas eng wks. Heat supply facility | Gas: 40,000Nm³/hr |

**Table 7.5**   continued

| Date enacted: | Nagoya<br>Feb. 1979 | Hyogo<br>Mar. 1979 | Mie<br>Mar. 1979 |
|---|---|---|---|
| Roads<br>national expressways | | | |
| others | 4 lanes/1km | 4 lanes/10km | 4 lanes/10km<br>(2 lanes/1km) |
| Dams<br>Lakes<br>Rivers | | 200ha | Storage 3 million $m^3$<br>hgt.30m/Lth.300m |
| Railways | Not specified | | Bullet train |
| Airports | | | Not specified |
| Land reclamation<br>Dumping | 10ha | PWAR Law | 15ha |
| Soil extraction | | | |
| New residential | 1,000 houses | 100ha | 100(10)ha |
| Industrial estate | 3ha | 100ha | 50(10)ha |
| New urban infrastructure<br>Distribution business centres<br>HUDC housing development<br>JRDC land preparation<br>EPSC land creation<br>Land creation for<br>  agricultural use | 10ha | Not specified | 100(10)ha |
| Housing redevelopment<br>Factories | 50ha<br>Site preparation: 9,000$m^3$<br>Construction: 3,000$m^3$ | Not specified | |
| Ports and harbours | | | |
| Waste treatment plants<br>garbage<br>sludge<br>waste disposal (fill) | 200T/d<br><br>3ha | 450T/d<br>150kl/d<br>15ha | <br><br>10(2)ha |
| Water supply<br>Sewage | Not specified | WW 10,000m3/d<br>100,000 people | sewer lines |
| Recreation projects | | 100ha | 50(10)ha |
| Power plants<br>hydro<br>thermal<br>nuclear<br>geotherm | 5,000kW | 30,000kW<br>75,000kW<br>all<br>10,000kW | 30,000kW<br>150,000kW<br>other |
| Other | Electricity related<br>facilities<br>Other | Oil: 15kl/hr<br>High buildings: 60m<br>Gas production:<br>500,000$m^3$/d | Livestock: 75,000$m^2$<br>Other |

| Date enacted: | Yokohama Jan. 1980 | Chiba Dec. 1980 | Saitama Feb. 1981 | Shiga Mar. 1981 |
|---|---|---|---|---|
| Roads national expressways | | | | |
| others | 4 lanes/3km | 4 lanes/10km | 4 lanes/10km | 4 lanes/10km (2 lanes/2km or 5mWth) |
| Dams Lakes Rivers | | 200ha 100ha | 200(30)ha | 50ha 20ha |
| Railways | Not specified | Bullet train | Not specified | |
| Airports | | Not specified | Runway: 2,500m Extension: 500m | |
| Land reclamation Dumping | 15ha | 50ha | | 3ha |
| Soil extraction | | 100ha | | 20(10)ha |
| New residential | 50ha | 50ha | 50(20)ha | 20ha |
| Industrial estate | 10ha | 100ha | 30ha | 20ha |
| New urban infrastructure Distribution business centres HUDC housing development JRDC land preparation EPSC land creation Land creation for agricultural use | 10ha | 100ha | 20ha | |
| Housing redevelopment Factories | 40ha | 100ha | 50ha | 40ha 10ha |
| Ports and harbours | | 300ha | | Not specified |
| Waste treatment plants garbage sludge waste disposal (fill) | 200T/d 5(2)ha | 450T/d 250kl/d 10ha | 200T/d 250kl/d 10ha | 100T/d 100kl/d 5ha |
| Water supply Sewage | WW 20,000m³/d Not specified | WW 10,000m³/d 15ha (200,000 people) | 5,000m³/d 20ha (200,000 people) | 2,000m³/d 5ha |
| Recreation projects | | 100ha | 50(20)ha | 20(10)ha |
| Power plants hydro thermal nuclear geotherm | | 30,000kW 150,000kW all | | |
| Other | Gas: 40,000Nm³/hr | Incinerator 450T/d Drainage: 100ha Other | Gas 40,000Nm³/hr Other | Gas 40,000Nm³/hr Drainage: 20ha Other |

## Table 7.5  continued

| Date enacted: | Hiroshima (Hiroshima City) Dec. 1982 | Ibaraki Apr. 1983 | Kagawa Sep. 1983 |
|---|---|---|---|
| **Roads** national expressways | | | |
| others | 4 lanes/10km | 4 lanes/10km | 4 lanes/10km |
| **Dams** Lakes Rivers | 200ha | | 200ha |
| Railways | Not specified | | Not specified |
| Airports | Runway: 2,500m Extension: 500m | | Runway: 2,500m |
| **Land reclamation** Dumping | 50(15)ha | 50ha | 50(15)ha |
| Soil extraction | | 50ha | 20ha |
| New residential | 100ha | 100ha | 30ha |
| Industrial estate | 50ha | 100ha | 20ha |
| New urban infrastructure Distribution business centres HUDC housing development JRDC land preparation EPSC land creation Land creation for agricultural use | 100ha | | 20ha |
| Housing redevelopment Factories | 50ha | 100ha | 100ha 20ha |
| Ports and harbours | | | |
| Waste treatment plants garbage sludge waste disposal (fill) | 300T/d 150kl/d 15ha | 300T/d 300kl/d 10ha | 300T/d 250T/d 30ha |
| Water supply Sewage | WW 10,000m$^3$/d 20ha | 200,000 people | WW 10,000m$^3$/d 150,000 people |
| Recreation projects | 50ha | | 20ha |
| Power plants hydro thermal nuclear geotherm | | 30,000kW 150,000kW all | 20ha Gas fired: 10,000Nm$^3$ |
| Other | Oil: 15kl/hr Other | Other | Gas: 100,000Nm$^3$/hr Gas prod: 20ha Other |

| Date enacted: | Nagano<br>Jan. 1984 | Osaka (Osaka City)<br>Feb. 1984 | Aichi<br>Mar. 1986 | Kyoto<br>May 1989 |
|---|---|---|---|---|
| **Roads** | | | | |
| national expressways | | | | all |
| others | 4 lanes/10km<br>(2 lanes or 5mWth) | 4 lanes/5km | 4 lanes/10km | 4 lanes/10km |
| **Dams** | | | | |
| Lakes | 100(30)ha | 100ha | 200ha | 200ha |
| Rivers | | | | 100ha |
| Railways | Bullet train | 3km | 10km | all |
| Airports | Runway: 2,500m<br>Extension: 500m | Not specified | Runway: 2,500m | Runway 2,500m<br>Extension: 500m |
| **Land reclamation** | | | | |
| Dumping | | 50ha | 50(30)ha | 50ha |
| Soil extraction | | 20ha | | |
| New residential | 20ha | 100ha | 100ha | 100ha |
| Industrial estate | 65ha | 50ha | 100ha | 100ha |
| New urban infrastructure | | 100ha | | 100ha |
| Distribution business centres | 20ha | 50ha | 100ha | 100ha |
| HUDC housing development | | | | |
| JRDC land preparation | | | | |
| EPSC land creation | | | | |
| Land creation for | | | | |
| agricultural use | | 500ha | | 500ha |
| Housing redevelopment | | 100ha | 100ha | 100ha |
| Factories | | | | |
| Ports and harbours | | | | |
| **Waste treatment plants** | | | | |
| garbage | 200T/d | 200T/d | 200T/d | 200T/d |
| sludge | 250T/d | 100kl/d | 200kl/d | 100kl/d |
| waste disposal (fill) | | 10ha | 30ha | 30ha |
| Water supply | | WW 10,000m$^3$/d | WW 10,000m$^3$/d | WW 10,000m$^3$/d |
| Sewage | 15ha | 100,000 people | 15ha | |
| Recreation projects | Golf 80ha | | 100ha | 100ha |
| **Power plants** | | | | |
| hydro | | 30,000kW | 30,000kW | 30,000kW |
| thermal | | 150,000kW | 150,000kW | 150,000kW |
| nuclear | | all | | all |
| geotherm | | | | |
| Other | Villas: 100ha<br>Ski: 30ha<br>Other | Gas: 40,000Nm$^3$/hr<br>Other | Oil: 15T/hr<br>Drainage<br>channel: 100ha<br>Other | Oil: 15T/hr<br>Land readjust-<br>ment |

## Note to Table 7.5

*Key:* () = smallest scale of development requiring EIA depending on local conditions; HUDC – Housing and Urban Development Corporation; JRDC – Japan Regional Development Corporation; EPSC – Environmental Pollution Service Corporation; d – day; h – hour; Hgt – height; L – length; Wth – width; kl – kilolitres; kW – kilowatts; T – tonnes; WW – waste water.

Sources: Kajima Construction Co. (1987), pp. 24–5; Shimazu (1987); Environment Agency (1985), pp. 43–8.

the industry's response to the first round of meetings and comments. The number of meetings varies according to the project, but two surveys from mid-1987 give an indication. In Tokyo, of forty-two projects, eleven required one meeting, six required two and the rest required more – up to fourteen in one case: on average, about four meetings were held. In Kawasaki, of forty-four projects, two required one meeting, twenty-eight required two and the rest required between three and five meetings; on average, two to three meetings were held.[6]

Public hearings are mandatory in Tokyo and Saitama. Other local governments' procedures make public hearings dependent on whether the governor or the review committee want more information.[7]

### Administrative review

The interim group's administrative review procedures require only that the directors of concerned government departments comment on the proposed project plan, and that these comments be taken into account when the project is implemented. However, the extensive and uniform groups (except Hokkaido) require that the governor give his opinion of the EIS based on citizens' comments, the comments of a review committee, and possibly public hearings.

The administrative review committee is usually composed of fifteen to twenty specialists: lawyers, professors and engineers. Only Kawasaki allows local residents also to be on the committee. In some cases, the committee's opinion can be inspected by the public, but in most cases it simply forms the basis for the governor's opinion.[8]

### Procedures of other EIA-related laws

Prior to the Cabinet decision, EIA procedures were already included in the legal requirements for

— factory siting (1973),
— reclamation of public water areas (1973),
— development in the Seto Inland Sea (1973),

**Table 7.6** Impacts assessed in local authority EIA procedures (national included for comparative purposes)

| | National | Kawasaki | Hokkaido | Tokyo | Kanagawa | Kobe | Okayama | Nagoya | Hyogo | Mie | Yokohama | Nagasaki |
|---|---|---|---|---|---|---|---|---|---|---|---|---|
| *Pollution control* | | | | | | | | | | | | |
| Air pollution | * | * | * | * | * | * | * | * | * | * | * | * |
| Water pollution | * | * | * | *+gw | * | * | * | * | * | *+gw | * | * |
| Soil pollution | * | * | * | * | * | * | * | * | * | * | * | * |
| Noise | * | * | * | * | * | * | * | * | * | * | * | * |
| Vibration | * | * | * | * | * | * | * | * | * | * | * | * |
| Offensive odour | * | * | * | * | * | * | * | * | * | * | * | * |
| Low frequency vibration | | | | | * | # | | # | * | * | | |
| *Living environment* | | | | | | | | | | | | |
| Electrical interference | | * | | * | * | * | | * | # | * | * | * |
| Sunshine impediment | | * | | | * | | | * | # | | | |
| Waste matter | | * | | | * | * | * | * | * | * | * | |
| Climate | | | | | | | | | | | | |
| Topography | * | * | * | * | * | # | * | * | | * | * | * |
| Geology | * | * | * | * | * | # | * | * | | * | * | * |
| Ground subsidence | | * (f) | | * | * | # | | * | | | | |
| Marine climate | | | * | | * | | | | * | * | * | * |
| *Natural environment* | | | | | | | | | | | | |
| Flora | * | * | * | * | * | * | * | * | * | * | * | * |
| Fauna | * | * | * | * | * | * | * | * | * | * | * | * |
| Marine life | | * | | * | | | | | | | | |
| Ecosystem | | | | | * | | | | | | | |
| *Socio-cultural environment* | | | | | | | | | | | | |
| Cultural assets | | * | # | * | * | * | * | * | * | * | * | * |
| Local community | | * | | * | * | | | | | * | * | |
| Scenery | * | * | * | * | * | * | * | * | * | * | * | |
| Outdoor recreation | * | | | * | * | | | | * | | * | |
| Accidents/safety | | # | # | | # | | | * | | | | * |

Notes: * = assessment required; # = if necessary; +gw = and ground water; (f) = factories

Source: Environment Agency (1985); Shimazu (1987); Kyoto Prefecture (1989).

**Table 7.6** continued

| | Chiba | Saitama | Shiga | Hiroshima | Ibaraki | Kagawa | Nagano | Osaka | Aichi | Kyoto |
|---|---|---|---|---|---|---|---|---|---|---|
| *Pollution control* | | | | | | | | | | |
| Air pollution | * | * | * | * | * | * | * | * | * | * |
| Water pollution | * | *+gw | *+gw | * | * | * | * | * | * | * |
| Soil pollution | * | * | * | * | * | * | * | * | * | * |
| Noise | * | * | * | * | * | * | * | * | * | * |
| Vibration | * | * | * | * | * | * | * | * | * | * |
| Offensive odour | * | * | * | * | * | * | * | * | * | * |
| Low frequency vibration | * | | * | * | * | * | * | * | * | * |
| | | | | | | | | | | |
| *Living environment* | | | | | | | | | | |
| Electrical interference | | | | | | | | * | | |
| Sunshine impediment | | | | | | | | * | | |
| Waste matter | | | | | | | | | | |
| Climate | | | | | | | | | | |
| Topography | * | * | * | * | * | * | * | * | * | |
| Geology | * | * | * | * | * | * | * | * | * | |
| Ground subsidence | * | * | * | * | * | * | * | * | * | |
| Marine climate | | | * | * | | | | * | | |
| | | | | | | | | | | |
| *Natural environment* | | | | | | | | | | |
| Flora | * | * | * | * | * | * | * | * | * | |
| Fauna | * | * | * | * | * | * | * | * | * | |
| Marine life | | | | | | | | | | |
| Ecosystem | | | | | | | | | | |
| | | | | | | | | | | |
| *Socio-cultural environment* | | | | | | | | | | |
| Cultural assets | | | | | | | | * | | |
| Local community | | | | | | | | | | |
| Scenery | * | * | * | * | * | * | * | * | * | |
| Outdoor recreation | | | | | | | | * | | |
| Accidents/safety | | | | | | | | * | * | |

Note: Information on all impacts assessed according to Kyoto Prefecture's EIA procedures was not available at the time of writing.

— port and harbour construction (1974), and
— development in natural park (1975).

In addition, ministries developed guidelines for

— power station siting (MITI: 1973, 1979),
— bullet-train lines (MoT: 1979), and
— city planning (MoC: 1985).

This section summarizes the EIA procedures required by these laws and guidelines (with the exception of bullet-train lines for which no information was available at the time of writing).

*Factory siting*

In October 1973 the Factory Location Law was amended to include provisions for EIA as shown in Figure 7.2. When determining factory

**Figure 7.2**  EIA procedure for factory siting

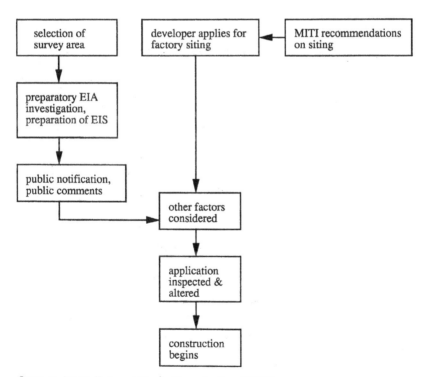

Source: Japan External Trade Organization (1983).

location MITI and related local authorities must:

— determine the area to be surveyed, namely the proposed site and surroundings;
— collect background data on, e.g., meteorology, hydrology, topography, social factors, land use, environmental protection plans, development plans;

The factory developer must then produce a report:

— describing the layout, effluent volume and flue gas volume of the proposed factory;
— describe environmental conditions at the site: currents, tides, water quality, biological conditions, wind direction/speed, and $SO_x$ and $NO_x$ concentrations;
— predict pollution levels by numerical simulation; and
— make the results known to other local enterprises.[9]

The resulting information is considered by MITI and the local authorities along with the application for construction. There are no procedures for public participation.

*Reclamation of public water areas*

The Cabinet decision includes procedures for assessing reclamation projects of ≥50ha in public water areas. However, the impact of reclamation projects had been assessed since September 1973, when the Public Water Areas Reclamation Law (PWARL) was amended in accordance with the 1972 Cabinet decision to include provisions for EIA and environmental protection measures when reclamation required licensing.

Figure 7.3 summarizes the PWARL's EIA procedures. When developers apply to a prefectural licensing authority for permission to reclaim land, they must submit with the application

— a report outlining the project;
— a land use plan;
— a financial plan;
— an agreement on rights of access for the public water areas concerned; and
— a report on planned environmental protection measures (an EIS).

The EIS must consider the environmental and social effects of the actual reclamation and of the post-reclamation use. It must:

— discuss existing environmental conditions at the proposed reclamation site and its surroundings;
— justify the selection of items for assessment;
— propose environmental protection measures;

**Figure 7.3**   EIA procedure for reclamation in public water areas

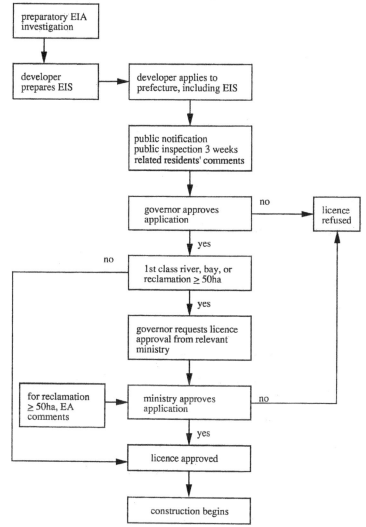

Source: Kajima Construction Company (1987), p. 40.

— discuss anticipated effects on the environment of the reclamation work
   and proposed environmental protection measures; and
— evaluate the project.[10]

The prefectural governor releases these documents for public inspection for three weeks and requests the opinions of interested parties. The

governor then considers these opinions when making a decision. For reclamation in bays or major rivers, or for reclamation of ≥50ha, the governor must get approval from the appropriate ministry; for development of ≥50ha the ministry must also consult the EA.[11]

Several exceptions and additions to the PWARL exist:

— According to the national EIA guidelines, the MHW, MoC, MoT and MAFF assess reclamation works which are under their jurisdiction.[12] These include disposal sites for domestic and industrial wastes of ≥30ha, and MAFF reclamation and drainage for land improvement. In these cases, the ministry carries out an EIA using both the national guidelines and the PWARL's procedures.[13]

— An application for a reclamation licence for the Seto Inland Sea must include a more comprehensive EIA (see next section).

— Twenty-two local authorities have their own EIA procedures for reclamation projects as small as 3ha. When the project scale is ≥50ha, local authorities must follow both their own procedures and those of the PWARL.

### Development in the Seto Inland Sea

The Seto Inland Sea requires special protection because it is a natural park and already heavily polluted. The Interim Law for Conservation of the Environment of the Seto Inland Sea of 1973 requires that any development in the Seto Inland Sea must be preceded by studies to consider the protection of the marine environment, natural environment, marine products and resources.[14]

### Port and harbour construction

The Port and Harbour Law was amended in 1973 to include EIA procedures; these are summarized in Figure 7.4. In 1974 the MoT released guidelines and standards detailing items to be considered by port authorities before they undertake major port developments; basically, these items are port usage, environmental effects and protection of shipping lanes. When a port or harbour development is proposed, the port authority must draw up a draft plan based on the MoT guidelines, and prepare a report on the development's environmental and social effects and proposed environmental protection measures.

The law does not require local consultation. However, the port authorities generally consult informally with fishermen, local authorities and other interested parties. A port study group composed of specialists and representatives of local interest groups is normally set up to comment on the plan. The MoT must be consulted, and in turn the MoT requests

**Figure 7.4**   EIA procedure for port and harbour construction

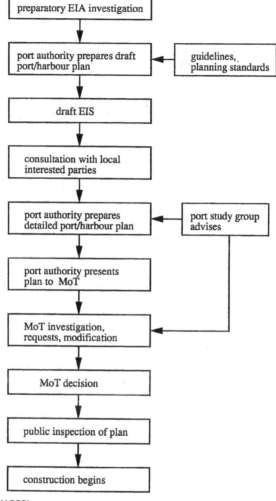

Source: Oshima (1989).

the opinions of the port study group and the EA. The MoT then gives the port authority notice of its decision and the plan is made open for public inspection.[15]

*Development in natural parks*

Under the 1975 amendments to the Natural Parks Law, permission from the governor or EA is required before large-scale projects can be built in special areas of national or quasi-national parks (see Chapter 5, 'Nature Conservation') and their environmental impacts must be studied. Procedures for these investigations are given in 'Investigation guidelines and application procedures for major developments in natural parks (excluding ordinary areas)'. The following projects require assessment:

— projects of ≥1ha, excluding new roads and developments for fishing and agriculture;
— new roads of ≥2km length or ≥10m width, except if they are approved in the Natural Parks Law or if specified conditions exist;
— any other activity that may have a significant environmental impact.

When applying for permission to develop, the developer must present the licensing authority with an EIS which addresses

— park use around the proposed project;
— the development's effect on flora, fauna and the area's scenic quality;
— the environmental and socio-economic usefulness of the development;
— proposed measures to decrease the predicted environmental impact; and
— alternatives to the existing development proposal which would accomplish the same purpose, and comparative studies of the environmental impacts of the proposed development and the alternatives.

These EIA procedures are the strictest in Japan. Although they do not include public participation procedures, they otherwise represent a near-ideal system which it would be wise to adopt at the national level.[16]

*Power station siting*

Electric power production in Japan was privatized in 1951, and is broadly controlled by MITI. The Science and Technology Agency is responsible for nuclear fuel reprocessing facilities and reactors for research and development.

Power stations are not covered by the 1984 national EIA guidelines. The LDP is sensitive to the views of the electric power companies, and has allowed them more freedom than other industrial sectors because energy production is considered a matter of national importance. In the negotiations leading up to the Cabinet decision, the electric power companies and MITI repeatedly argued that it would be difficult to implement the proposed EIA measures and that the public consultation procedures involved would delay Japan's energy programme.

Instead, MITI developed its own EIA procedures. In December 1973 it began requiring electric power companies to conduct EIAs for proposed power stations, which it reviewed when making siting decisions. On 4 July 1977 MITI resolved to strengthen these procedures, and on 26 June 1979, MITI's Agency for Natural Resources and Energy[17] released

— guidelines for conducting EIA for power stations;
— guidelines for public participation in environmental reviews;
— guidelines for conducting environmental reviews; and
— guidelines for how to use the other guidelines.[18]

These guidelines apply to thermal power plants of ≥150,000kw, geothermal power plants of ≥10,000kw, hydropower plants of ≥30,000kw, all nuclear power plants, and other power plants considered to need an EIA. In those cases where both local regulations and MITI procedures apply, both are used. No special procedures are required for nuclear power plants. EIAs for commercial nuclear fuel enrichment and reprocessing facilities are expected to be required by the 1990s when these facilities are scheduled to begin operation. EIAs for nuclear waste disposal facilities are not yet being considered.[19]

MITI's EIA procedures for power plants are summarized in Figure 7.5. According to MITI's guidelines for conducting EIAs, an electric power company's EIS must include

— an outline plan of the power station;
— a review of current environmental conditions at the proposed site;
— proposed measures for environmental protection;
— an estimate and assessment of environmental impacts; and
— an evaluation of the development project.[20]

The EIA must discuss the types of impact listed in the national guidelines (see Table 7.3), minus odours and recreation, plus climate, sea floor topography, marine life, marine geology, and social, economic and cultural factors. The detailed consideration of impacts on the terrestrial and marine ecosystems and on the marine and fishing industry is notable. Radioactivity and risk assessment are not considered in the EIA, but instead are dealt with by the Atomic Energy Commission and the Nuclear Safety Commission.

MITI's guidelines for public participation in environmental reviews require that the electric power company release an EIS to the public for twenty days. During that time, it must distribute information on the project and its environmental implications, and hold a public hearing. For thirty days after the release of the report, the company should accept citizen comments, and then present pertinent comments and its replies to MITI in its conservation plan.[21]

The guidelines for environmental review specify that MITI must then

— review the project plan and current environmental conditions;

**Figure 7.5** EIA procedure for siting power stations

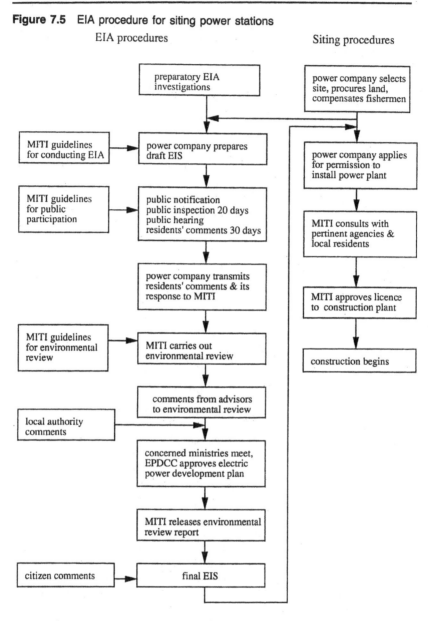

EIA procedures                                    Siting procedures

preparatory EIA
investigations

power company selects
site, procures land,
compensates fishermen

MITI guidelines
for conducting EIA → power company prepares
draft EIS

power company applies
for permission to
install power plant

MITI guidelines
for public
participation → public notification
public inspection 20 days
public hearing
residents' comments 30 days

MITI consults with
pertinent agencies &
local residents

power company transmits
residents' comments & its
response to MITI

MITI approves licence
to construction plant

MITI guidelines
for environmental
review → MITI carries out
environmental review

construction begins

comments from advisors
to environmental review

local authority
comments

concerned ministries meet,
EPDCC approves electric
power development plan

MITI releases environmental
review report

citizen comments → final EIS

Source: Kurihara *et al.* (1982), p. 292.

— select items for examination;
— examine possible environmental protection measures;
— examine the EIS submitted;
— examine the adequacy of the proposed environmental monitoring
  system; and
— comprehensively assess the proposed project.[22]

MITI then sends a draft environmental examination report to the Electric
Power Development Coordination Council,[23] revises its report based on
the EPDCC's comments, and makes its final environmental report public.
After these EIA procedures, the electric power company can apply to
MITI for a licence to construct and operate a power station; MITI in turn
holds public hearings and consults with other agencies before determining
whether the permission will be granted or not.[24]

Between 1977 and 1984 MITI EIAs were carried out for thirty-four
thermal, one geothermal, fourteen hydropower and eleven nuclear power
stations.[25]

*City planning*

In April 1985 the MoC's City Planning Bureau released the guidelines
'Implementation of EIA for Developments under MoC Jurisdiction' in
accordance with the 1984 Cabinet decision. Under these guidelines, EIAs
for projects in city planning areas[26] differ from those in other areas in
that the enforcing authority is the city planning department rather than
the pollution control department (or whichever section is normally
responsible for supervising EIA procedures). In addition, EIA procedures for projects in city planning areas must conform to the
procedures specified in the City Planning Law.

Figure 7.6 shows the interrelationship of city planning and EIA
procedures. This relationship is not always clear, and confusion often
arises as to which procedures are applicable and which local authority
department is responsible for judging the assessment; this will be shown
in Chapter 13.

Article 32 of the City Planning Law (Administrative Consent for
Public Facilities) states that public consultation is required before a local
authority can permit a development which exceeds a specified standard
(i.e. as specified in Table 7.2). A draft EIS and a basic development plan
must be prepared before city planning procedures begin. Once a draft
plan is drawn up, the developer makes the plan and the EIS available for
two rounds of public inspection and comments, as shown in Figure 7.6.
The developer incorporates these comments into a final EIS. This EIS is
given to the licensing authority with the application for development
permission, and the necessary consultations take place. For projects in

**Figure 7.6** EIA procedure for city planning

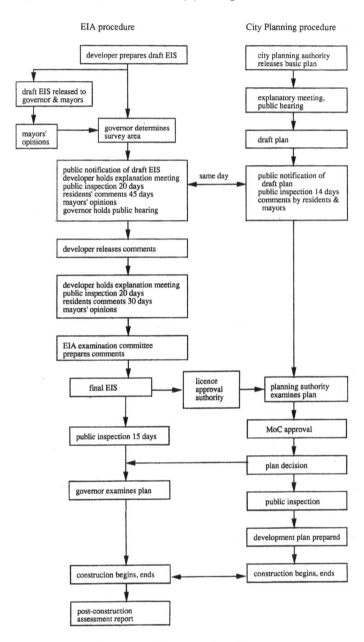

Source: Kajima Construction Company (1987), p. 43.

city planning areas which do not require assessment, approval takes about two months; when an EIA must be prepared, approval takes about 10.5 months.[27]

## Conclusions

This chapter has briefly described the EIA systems that exist in Japan. These systems vary widely in their procedures. Many overlap. They are administered with widely varying degrees of care and enthusiasm.

The different government bodies have drawn up EIA systems to suit their own administrative needs and goals, and the resulting EIA systems vary accordingly. The large ministries like MITI, MoC and MoT are mainly concerned with fostering economic growth and have developed EIA systems which consider environmental factors without causing unnecessary delay, increased cost or change of plans. Most local authorities, instead, instituted EIA systems in response to severe pollution problems or citizen pressure, and consequently their EIA procedures are quite stringent. The EIA system for natural parks, where environmental preservation is a primary consideration, is the only system which requires that alternatives to the proposed development be considered.

Essentially, the form which an EIA system takes depends on the priorities of the administration and the society as a whole. With minor exceptions, Japan's perceived priority is the pursuit of economic growth. An EIA system created under such circumstances will necessarily be weaker than one drawn up in a climate of strong environmental awareness. Without a powerful central body to coordinate environmental policies, one can expect that the existing non-uniformity and duplication of procedures will continue.

## Notes

1. Suzuki (1985), pp. 52–62.
2. If the explanatory meeting cannot be held for reasons not attributable to the developer (e.g. public protest), then it is not required. Alternatively, the developer can entrust the local authorities to hold these meetings.
3. The frequency of such events and the amount of compensation is still unclear: Environment Agency (Nov. 1984), p. 1.
4. Environment Agency (Nov. 1984).
5. Environment Agency (Mar. 1985), pp. 18–51.
6. Kajima Construction Co. (1987), p. 107.
7. Environment Agency (Mar. 1985), p. 21.
8. Ibid., pp. 34–5.
9. Japan External Trade Organization (1983), pp. 14–18.
10. Kajima Construction Co. (1987), pp. 41–2.
11. Ibid., pp. 40–1.

12. The MHW released EIA guidelines for reclamation projects in Dec. 1985, the MoC in April 1986, the MoT in March 1986, and the MAFF in Oct. 1986: Environment Agency (June 1987).
13. The national guidelines differ from the PWARL in that they do not require that the environmental impact of the *use* of the reclaimed land be examined: Kajima Construction Co. (1987), pp. 42–3.
14. Ibid., p. 42.
15. Oshima (1989).
16. Kajima Construction Co. (1987), pp. 43–7.
17. The 1977 report is entitled 'Strengthening of EIA and environmental review for the siting of power plants', and the 1979 report is 'Guideline for implementation of EIA and environmental examination of power plant siting': Kurihara *et al.* (1982), p. 193.
18. Ibid., pp. 291.
19. Ibid., p. 295.
20. Ibid., p. 293.
21. Ibid., pp. 293–4.
22. Ibid., p. 294.
23. The EPDCC is composed of representatives from various concerned ministries and agencies.
24. Japan Electric Power Information Centre (1986), p. 27.
25. Kajima Construction Co. (1987), p. 37.
26. In 1985 designated city planning areas covered about 92,000km$^2$, 24 per cent of the national land area.
27. Ibid., p. 38.

# Chapter eight

# Critical discussion of environmental impact assessment in Japan

The development of environmental policies is moulded by the inter-actions of the government, industry and other interest groups.[1] These in turn are influenced by a broad array of physical, socio-cultural and economic factors. In the traditional Japanese view, nature and humans are not separate entities, and environmental preservation is not 'right' in the same sense as Westerners view it.[2] Thus an essentially Western concept like EIA, which emphasizes protecting the natural environment from human destruction, could never be simply imported unchanged in Japan.[3] When EIA was adopted into Japan's decision-making structure it was 'Japanized' so that some of the elements that in the West are considered a prerequisite for its effective implementation were severely weakened. The result is a strange hybrid: judged by Western standards, it is intrinsically weak, and from a Japanese perspective it conforms badly with the traditional decision-making process.

This chapter reviews some strengths and weaknesses of Japan's EIA system. Such a judgement must be preceded by a position statement: we judge EIA systems by whether they accomplish the goal of preventing environmental harm. An ideal EIA system would:

— apply to all projects that are expected to have a significant environmental impact, and address all impacts that are expected to be significant;
— compare alternatives to a proposed project (including the possibility of not developing), management techniques, and mitigation measures;
— result in a clear EIS which conveys the importance of the likely impacts and their specific characteristics to non-experts as well as experts in the field;
— include broad public participation and stringent administrative review procedures;
— be timed so as to provide information for decision-making;
— be enforceable; and
— include monitoring and feedback procedures.

It's a tall order. No perfect EIA system exists, and any country's EIA system will be both strengthened and weakened by the milieu in which it is applied.

This chapter discusses the following points:

— legal status
— projects subject to EIA
— scope of the EIA
— timing and alternatives
— technical standards
— public participation
— administrative review
— and other factors.

### Legal status

With the exception of those local authorities with EIA ordinances and several laws which include mandatory EIA provisions, EIA in Japan does not have a legal basis. This is partly explained by the Japanese preference for administrative guidance and non-interventionism. Japan's quasi-elitist decision-making structures and the close relationship between industry and government are also influential. This lack of legal status means that neither the correct execution of EIA procedures (procedural duty) nor the EIA's adequacy (substantive duty) can be challenged in court and enforced.

### Projects subject to EIA

The list of projects subject to the Cabinet decision is severely limited. One way to show this is to compare the list with that of the EC directive on Environmental Assessment. The following types of development, considered in the EC Directive to have a significant environmental impact, are not subject to Japan's national EIA procedures:

— trading ports
— crude oil refineries
— integrated chemical installations
— integrated works for the initial melting of cast-iron and steel
— installations for extracting and processing asbestos
— thermal and nuclear power stations,
— and installations for storing and disposing of radioactive waste.

In addition, the EC directive lists eighty types of development which may have significant environmental effects and which should be assessed if the member state considers that their characteristics so require; only eight of these are covered by Japan's national guidelines. The Cabinet decision does not expand national jurisdiction over environmental issues. It is applied through the permit procedures set up by currently existing laws, and does not extend to new projects not currently requiring authorization by the national government.[4]

Japan's local authority impact systems tend to be more comprehensive than those at the national level. For instance, the extensive and uniform groups' EIA procedures generally apply to:

— recreation facilities
— waste treatment facilities
— ports and harbours
— factories
— distribution centres
— redevelopment projects
— and power plants.

In Japan, as in other countries, there is little examination of the environmental consequences of policy options. Assessment takes place at the project stage of the planning process; at this stage alternative policy options have long been discarded. Local authorities, however, are beginning to institute a system of environmental management based on the concept of carrying capacity, which integrates REMPs, EIA and environmental monitoring. A further development would be a two-tier arrangement for assessing major resource projects: in the first stage policies would be assessed using such criteria as aims, demand forecasts, safety and environmental protection, and in the second stage the repercussions of individual projects would be assessed.[5]

*Scope of the EIA*

Not included in Japan's national guidelines are the consideration of accidents, radioactivity, impediments to television and sunshine reception, cultural heritage, the maintenance of mixed natural/man-made scenery, or long-term and cumulative effects of pollution. Radioactivity, sunshine and cultural heritage were excluded because they are regulated by other laws; accidents and scenery were felt to be already adequately considered.[6] Some of these omissions can also be explained by Japan's greater concern with human health impacts and pollution control than with the natural environment.

Socio-economic factors must be described in Japan, but their in-depth assessment is thought unnecessary. Techniques for socio-economic impact assessment are not well developed and the government is unwilling to open such important matters to assessment when the results cannot be trusted. Furthermore such an assessment is very politically sensitive, since socio-economic effects are closely related to costs, benefits and their distribution. Many researchers and administrators are unwilling to involve themselves in what could easily turn into highly political issues. At the 1982 Diet discussion, for instance, it was argued that socio-economic impact assessments might give local residents too much control over the acceptance or rejection of projects.[7] It has also been argued that the role of EIA is to look only at negative aspects of development, and that economic effects are usually positive and thus do not require assessment. However, this is an invalid assumption and results in an incomplete assessment.

*Timing and alternatives*

The timing of EIA *vis-à-vis* major development decisions (e.g. location, size) determines how the information from an EIA is used. Obviously timing is integrally linked with the consideration of alternatives. In Japan alternative sites, designs or even mitigation methods do not have to be considered. Without such a requirement, a developer has no incentive to consider less environmentally damaging developments or management methods, or the possibility of not developing. If developers need to consider only one plan, they will first choose the project's location, scale, etc., and only then prepare an EIA. Thus EIA becomes used at the end of, rather than during, decision-making, and its timing becomes too late. EIA then often merely justifies development decisions and attempts overcome local opposition.

A mandate to study alternatives was included in early drafts of the Cabinet decision, but was gradually weakened and finally removed around 1980.[8] Several explanations have been given for this lack of an alternatives requirement:

— The strong factionalism and carefully defined jurisdictions of government planning bodies make it difficult for one planning body to consider alternatives which might cause another body to change its plans.[9]
— The scarcity of developable land limits the number of possible alternative sites for a project. Likewise, due to high land prices the project site is a primary consideration for the developer, and the government is sympathetic to this problem.[10]
— Instituting an alternatives requirement would involve choosing one alternative and rejecting the others. This contradicts Japan's traditional 'do not produce a loser' consensus system. In the past, when several alternatives existed, all of them were chosen: for instance, three Honshu–Shikoku bridge routes are being built although a need for only one has been demonstrated, because rejection of a route

would offend the people who support that route (see Chapter 10).
— Finally, consideration of alternatives is expensive and therefore
 opposed by commercial and industrial interests.

Local authorities also do not require alternatives to be considered.
One reason for this is that an alternatives requirement could be
economically harmful to the authority. Without a unified national EIA
system, developers are likely to build in areas of least resistance,
'pollution havens'. If local authorities forced developers to consider
alternatives, the developers might decide not to develop in that area. In
addition, the local authorities might not be able to afford to consider
alternatives to their own proposed projects.

Figure 8.1 is taken from a recent book on EIA by a major Japanese
construction company. The cycle of 'plan – protection measure – new
plan' assumes that all environmental problems can be solved simply by
adding enough pollution control measures. This implies that EIA should
not seek to limit inappropriate development but should merely indicate
suitable pollution control measures. This shows just how little the Cabinet
decision has expanded the scope of Japanese environmental policy. The
lack of consideration of alternatives is indicative of Japan's views of the
roles of economic development and environmental policy: development is
considered necessary; environmental policy merely counteracts its
negative effects.

**Figure 8.1**  Industry view of EIA procedure

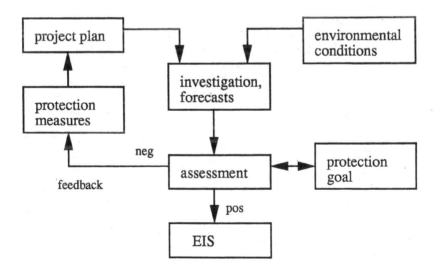

*Technical standards*

Japan's EIA system is greatly strengthened by its high degree of technical sophistication. Japan has very advanced systems of data collection, analysis and simulation. Computer simulation is particularly advanced. Environmental monitoring is also well established (see Chapter 5), although the data gathered often seem to be used only to ensure that EIA predictions are not exceeded, not to improve methodology.

*Public participation*

The lateness of Japan's economic and political development compared to that of Western countries is often used to excuse its relative lack of democratic principles. Whereas in the West EIA is seen as part of the development towards more environmentally and socially conscious decision-making, in Japan this concept remains in its infancy. Public involvement in administrative decision-making is not encouraged in Japan, and the national EIA procedures do not allow for broad public participation. One politician, during the 1982 Diet discussions of the EIA bill, went so far as to question whether Japan's political climate was prepared for the 'direct democratic principles' involved in citizen participation.[11]

In the discussions leading up to the EIA bill's introduction to the Diet, the role of public participation in EIA was defined not as helping to set policy, or even stating the opinions of interest groups, but merely providing information to decision-makers and ensuring that all issues are considered. During the 1982 Diet debate about the EIA bill, an EA spokesman noted that 'citizens' opinions should be heard, but they should not directly participate in the industry's decisions'.[12]

According to the Cabinet decision, the definition of concerned citizens is determined by the developer (in consultation with local authorities) and need only include those citizens living in areas that will be directly affected by the proposed development. Environmental groups and specialists from outside the area are not allowed to comment, since their opinions are not thought to reflect the local social and environmental characteristics. Problems caused by such a narrow definition are illustrated in the case of the New Ishigaki Airport (Chapter 12).

The mode of public participation is also rather restricted. Residents can inspect and comment on the draft EIS, and the developer must explain the EIS at a public meeting, but the decision-making process is virtually closed.

Public hearings were originally mandated by the national EIA bill, but this clause was deleted in around 1979 for several reasons:

— development projects do not impinge on any citizen rights, so public hearings to discuss them were not felt to be needed;
— the range of views expressed at such meetings can cause confusion and detract from the most pertinent issues;
— hearings for national projects were considered unfeasible, and hearings for local projects are problematic because these projects are often influenced by national policies; and
— radicals might take over the meetings.[13]

Public hearings are a traditionally accepted practice in Japan, and as such are sometimes held for large national projects. However the hearings are often so formal and given at such short notice that people opposing the project cannot prepare for them adequately:[14] an example of this will be given in Chapter 14.

Local authorities are much closer and more vulnerable to public opinion, and consequently must make greater efforts to accommodate it. They generally have more open decision-making, more public involvement and more accountability. For instance, of the seventy-one EIAs completed by 1983 under local authority ordinances, forty-nine included arrangements for public participation and sixteen involved public hearings. Thirty cases took over one year, sixteen took six months to one year, and twenty-five took less than six months. There were no instances in which explanatory meetings could not start or were cancelled due to public protest, and there were no cases of litigation against a project based on its EIA.[15] These points are particularly important in that they undermine the arguments of those who oppose EIA on the grounds that it would foster public protest and delay.

*Administrative review*

The Cabinet decision specifies three types of administrative review: local authority, relevant national ministry and the EA. However, all three are open to bias.

Local governments rely heavily on national funding, and thus can be manipulated into approving the EISs of projects which the national government wants. In addition, local governments and national ministries are often both the project's developer and its EIA reviewer. Because of the close link between industry and the national government, even those projects that are not initiated by government bodies are often judged by ministries which are intrinsically in favour of economic development, and thus of the projects.

Review by the EA – the sole reviewer whose main concern is environmental protection – is only at the request of the concerned ministry, and only after the final EIS has been completed and made public. Since the EA's advice is unlikely to serve the best interest of the concerned ministry, the ministry is unlikely to ask for EA review. At this late stage, the EA certainly cannot suggest major changes in the project plans, and whatever input it does have is not binding on the ministry or the developer. The EA is also sensitive to its negative image among the ministries and tries to accommodate them. To date, it has approved all of the projects assessed under the Cabinet decision. It also approved all of the EISs based on the 1972 guidelines, although at times it did propose changes.[16] The New Ishigaki Airport (NIA) represents a rare case in which the EA made a firm stand against a development project (Chapter 12). The project, however, was not abandoned despite the EA's efforts.

The main role of Japan's present EIA system seems to be one of plan and policy coordination.[17] This function would be more attractive if Japan's policies were decided from the bottom up rather than from the top down. Many of the problems with Japan's environmental policies, particularly their seemingly inequitable nature, stem from the closed and hegemonic nature of the decision-making process. In the US, the Freedom of Information Act played a key role in opening up this process, but Japan has no equivalent although some local authorities have produced their own versions. EIA in Japan has not changed the balance of power in this process and is thus seen as merely a form of project justification.

*Other factors*

Japan's EIA procedures are unnecessarily complicated, non-uniform and duplicative. Eight laws contain EIA provisions, eight ministries and agencies are responsible for implementing and enforcing EIA procedures, and twenty-seven prefectures and designated cities follow separate local procedures. The range of developments subject to EIA, the scope of assessment and EIA procedures vary from one system to the next. Thus a project can be subject to different EIA procedures purely on the basis of geographical location. The Cabinet decision did not, as was hoped, foster the speedy implementation of an EIA system. It merely brought together a jumble of laws, guidelines, procedures and project types into one package. Coordination is urgently needed, but at the moment 'coordination' implies conformity with the national guidelines; this effectively hinders further action by local authorities and constitutes a serious weakening of the EIA systems (Chapter 13 – Trans-Tokyo Bay Highway).

The relative merits of EIA depend upon the perspective of the participants in the process. The developer wants certainty in the decision-making process, minimal administrative discretion, emphasis on technical assessment and market forces, and minimal delay and costs. The government wants control over the process and timescale of project approval, and limited public participation. Environmental groups want citizen control over development and increased expenditure on environmental protection.[18]

For the developer and government, Japan's EIA system ensures that environmental factors are considered in the decision-making process without undue delay and cost. The imposition of strict timescales with a 'go-ahead' default gives the developer much power and eases administrative responsibilities. In contrast the US system, with its minimum timescales and 'stop' default, leads to frequent and costly delays in project implementation, uncertainty about whether projects will be approved, and more administrative work.

For environmental groups, however, a system like that of the US would be preferable. Systematic processes such as scoping and tiering[19] lead to a more comprehensive analysis of a project's environmental impacts than negotiation and administrative guidance, which often seems like a bargaining process through which both sides can reach a mutual face-saving agreement rather than the most environmentally sound solution. In legal terms, mandatory EIA systems also offer more citizen control than guidelines.

Although Japan's system of EIA is relatively advanced by international standards, its performance is disappointing. The principle of impact assessment is fairly well established, but the practice is far from satisfactory. EIA in Japan is all too often used as a tool to justify development decisions and overcome local opposition. The consideration of long-term environmentally detrimental effects seems to be a side issue, as most impact studies tend to conclude that the project's environmental impact will be small.

### Notes

1. O'Riordan and Sewell (1981), p. 5.
2. The Japanese view of nature is discussed further in, e.g., Nakamura (1964), Moore (1967) and Koller (1985).
3. Bidwell (Sep. 1985).
4. Suzuki (1985).
5. O'Riordan and Sewell (1981)
6. Environment Agency (Nov. 1982), pp. 25–8.
7. Ibid.
8. The CCEPC's 1974 interim report proposed a 'comparative study of alternative plans'. Its 1975 report noted that 'consideration of alternative plans is desirable . . . [and] clarification of the process by which the original plan has been put into shape would be . . . beneficial'. The council's 1979 report stated that 'when the choice of alternatives is made prior to the

preparation of draft EIS, it will be appropriate to indicate . . . the alternatives already studied'. The 1984 Cabinet decision omits mention of alternatives altogether.

9. Isobe *et al.* (1979), pp. 20–1.
10. Environment Agency (May 1979).
11. Environment Agency (Nov. 1982), p. 47.
12. In the late 1970s, the EIA bill's word for 'participation' was changed from the more active *sanka* (participation, joining, entry) to the more passive *kanyo* (participation, being concerned with): Isobe *et al.* (1979), pp. 19, 26.
13. Ibid., pp. 26–7.
14. Hase (1981b), pp. 227–51.
15. Kajima Construction Co. (1987), pp. 27–36.
16. Interview with Iwata M., Environment Agency, Environmental Impact Assessment Division, July 1987.
17. Morita and Gotoh (1985).
18. Glasson and Elson (1987), pp. 39–40.
19. Scoping and tiering are concepts used in EIA in the US. Scoping is a negotiation process aimed at narrowing the topics considered in an EIA to those issues which are significant and not covered in previous EIAs. Tiering encourages agencies to eliminate repetitive discussions and to exclude from consideration those topics already decided and those not ripe for discussion.

Part three

# Case studies

# Chapter nine

# Overview of the case studies

## Introduction

In this part, we present and evaluate five case studies of EIA:

— Kansai International Airport
— Honshu–Shikoku Road/Rail Bridges
— New Ishigaki Airport
— Trans-Tokyo Bay Highway
— Kyoto Second Outer Circular Route.

These studies aim to provide insight into Japan's development process, practicalities of EIA preparation and problems encountered when environmental protection is seen to infringe on economic growth and development. Before presenting the case studies, however, we will briefly review the economic and social factors which have affected major development projects in Japan, and explain why the case studies were selected. The case studies are presented in Chapters 10–14.

## Major development projects in Japan

The siting of major projects poses serious environmental and social problems. Nowhere is this clearer than in countries like Japan where urban population densities are high and developable land is scarce. The history of project developments in Japan has been characterized by at times violent opposition to the government's plans. Narita airport near Tokyo, for example, was the scene of violent demonstrations in the 1960s in which a number of students and policemen were killed and hundreds were injured.

Major projects, however, are very attractive assets for any regional economy and have increasingly come to be viewed as a way to stimulate industrial activity. Consequently, Japan's development plans have concentrated on the development of new industrial/residential centres to act as growth poles in lieu of existing metropolitan areas, and the expansion of transport facilities to access and link Japan's less developed areas. Many of these projects were first proposed in the 1960s, but because of the long timescales involved in their planning and implementation, and in some cases the extent of public protest, many of them are only now under construction.

Another major stimulus for the renewed development of the Japanese archipelago was the appreciation of the yen from September 1985 onwards. This negatively affected Japan's export capability and forced a restructuring of the economy in the face of new competitive conditions. The government responded by pursuing a massive development programme to stimulate the domestic economy. It has been estimated that over the next few years about ¥20 trillion ($160 billion) will be allocated for public works and infrastructure projects.[1]

Several new airports are planned (see Figure 9.1), including ten airports capable of handling jets. One of these will be the 24-hour Kansai International Airport, which was originally proposed in the 1960s but the construction of which was delayed until 1987 by protest and funding problems. The KIA is scheduled for completion in 1993 and already there are plans for another 24-hour international airport on reclaimed land in Ise Bay near Nagoya.[2] The Tokyo International Airport will be re-sited in Tokyo Bay, and an additional runway will be built at the New Tokyo International Airport (Narita). The most controversial of the plans is that for the New Ishigaki Airport, the future of which is as yet uncertain due to concerns about its environmental impact.

The current 4,400km expressway system is expected to be expanded to 14,000km by early in the twenty-first century at an estimated cost of ¥10 trillion (see Figure 9.2). The new expressways will traverse Japan roughly horizontally, to supplement the existing system which runs primarily vertically, and will accommodate traffic of higher speeds than at present.[3] One of the expressway developments is an outer ring road around Tokyo similar to the M25 around London; one part of this is the Trans-Tokyo Bay Highway, which will cost ¥1.15 trillion. Railway expansion is also planned, with seventeen new lines proposed for the Shinkansen (bullet train), and the development of the Hikari Super Express, which will have a top speed of 300km/hr.[4]

The expansion of the transport network is supported by the construction of various bridges and tunnels to link Japan's main islands. When the three bridge links between Honshu and Shikoku are completed, they will include the world's longest, sixth-longest and twelfth-longest suspension bridges.[5] The 19km New Kammon Tunnel between Honshu and Kyushu was opened in 1975. It was the world's longest tunnel until 1987, when the 54km Seikan Tunnel between Honshu and Hokkaido was opened at an estimated cost of ¥850 billion ($3.4 billion).[6] Another tunnel is planned for the Kitan Strait between

**Figure 9.1** Airports

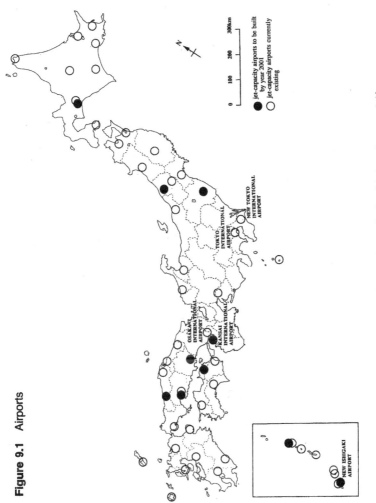

Source: *Newton* (special edn, February 1988), 'Japan in the 21st century: part 1', 3(8).

**Figure 9.2** Expressways

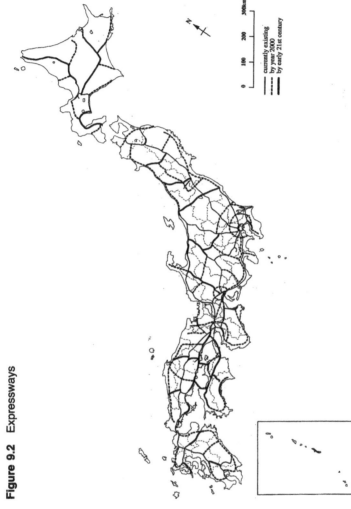

currently existing
by year 2000
by early 21st century

0   100   200   300km

Source: *Newton* (special edn, February 1988) 'Japan in the 21st century: part 1',
3(8).

Wakayama and Awaji Island; with the easternmost Honshu–Shikoku bridge link and existing highways, it will form a complete loop around Osaka Bay.

In addition, the private sector has embarked on massive development programmes, which include the construction of about twenty 'techno-polises' (see Chapter 4), several other theme-'polises', and dozens of major business and industrial projects. Table 9.1 and Figure 9.3 show some of these developments.

All in all, Japan has embarked on one of the world's most comprehensive plans for the further development of an industrialized nation. Although the feasibility and desirability of such plans is open to doubt, it is certain that they will have a major environmental impact.

**Table 9.1**   Major planned development projects

|  | Cost (in ¥1bn)* | Construction dates |
|---|---|---|
| 1. Kanjoumu greenbelt | | 1983– |
| 2. Aviation and space centre | | 1990– |
| 3. Magmapolis (geothermal centre) | 23 | 1980–90 |
| 4. Sanriku coastal marine zone plan | | 1984–90 |
| 5. Sendai international trade port | 138 | 1985–95 |
| 6. Soma area development enterprise | 400 | 1983– |
| 7. Makuhari new business district (437ha) | 1,000 | 1972–90 |
| 8. New government quarters | 136 | |
| 9. Saitama 'You and I' Plan | | 1984– |
| 10. Kawasaki Technopia | 80 | 1985–88 |
| Kanagawa Science Park | 65 | 1986–9 |
| 11. Minato–Mirai 21 (Yokohama waterfront reclamation/ development, 186ha) | 2,000 | 1983–2000 |
| 12. Shonan Nagisa shore plan | 250 | 1986– |
| 13. Soft Energy Model City | | 1986– |
| 14. Shimizu artificial island | | |
| 15. Marine Plan 21 | 50 | 1985– |
| 16. Atom-polis | | |
| 17. Kansai Academic City (11 areas, 16,000ha) | 1,200 | 1983–2000 |
| 18. N. Osaka International Cultural Park City | 1,000 | 1990–2000 |
| 19. Technoport Osaka | 2,000 | 1983–2005 |
| 20. Maejima (city linked to Kansai Internat. Airport, 318ha) | 140 | 1986–93 |
| 21. Wakayama Cosmopark extension | 150 | 1986–95 |
| 22. Rokko Island development (580ha) | 540 | 1972–90 |
| 23. Tokushima bulk distribution harbour | 45 | 1986–92 |
| 24. Nakaumi agricultural land reclamation | 88 | 1963–88 |
| 25. Iwakuni military base transfer offshore | 300 | 1986–90 |
| 26. Shimonoseki–Kokura artificial island | | 1986– |
| 27. Marinopolis | 119 | 1982–90 |
| 28. Osumi Biopolis | | 1984– |
| 29. Isahaya disaster prevention scheme | 135 | not available |

Note: * ¥1bn is approximately = $7.4 million, £4.2 million.
Source: *Newton* (special edn, Feb. 1988, in Japanese), 'Japan in the 21st Century: part 1', 3(8), pp. 90–1.

**Figure 9.3**   Development projects

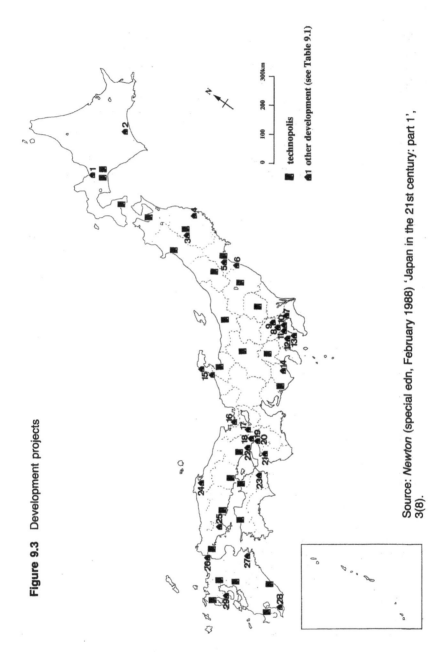

Source: *Newton* (special edn, February 1988) 'Japan in the 21st century: part 1',
3(8).

The Japanese have developed a number of strategies or tendencies for ensuring the successful implementation of their major projects. First, more and more projects are being built on reclaimed land in bay and coastal areas. According to the EA, from a total of 450 environmental assessments dealt with by local authorities between 1982 and 1986, 114 (25 per cent) were for reclamation projects, including seven for the construction of airports. Reclamation overcomes problems of land shortage, land ownership and public protest, although negotiations with fishermen's unions are necessary and compensation is often expensive. Moreover, the removal of land ownership issues reduces those vested interests that would normally work against the project, and increases the confusion of those opposed to the project. Only those residents directly facing the proposed development will oppose it, rather than all citizens in the vicinity. In this way public opposition to the project can be minimized. Two of the projects selected as case studies are for sites on reclaimed land: the Kansai International Airport and the New Ishigaki Airport.

Second, the environmental consequences of such large-scale projects are immense and adequate measures must be taken to minimize their adverse effects. EIA provides the developer and the administrative decision-makers with the opportunity to consider environmental factors more carefully. There is, however, little information about the effectiveness of these assessments. Many protest groups maintain that the present assessment system merely provides a means for the developer to diffuse local protest and push a project through the decision-making procedures smoothly. The Ishigaki airport and the Kyoto circular highway case studies in particular show how some local authorities seem to manipulate the EIA system to ensure that projects will go ahead. On the other hand, the authorities involved in the Kansai airport's assessment used the EIA procedures to obtain further environmental information and social benefits.

Third, three of the case studies involve the establishment of a mixed public/private sector authority to implement the project plan: the Honshu–Shikoku Bridge Authority, the Kansai International Airport Co., and the Trans-Tokyo Bay Highway Co. Although the creation of these organizations, which combine public sector management and private sector finance, is considered to be economically beneficial, they tend to undermine the workings of the democratic/protest system. Their boards of directors are not democratically elected and are thus not directly answerable to the public. This confuses protesters who do not know to whom to protest.

## Case study selection and format

The five development projects chosen as case studies of EIA in Japan were selected on the basis of how well they exemplify Japan's environmental, economic and social trends; the scale of environmental impacts involved; and the availability of information. The Ishigaki example was chosen because it is particularly contentious. The Kyoto circular highway will traverse Toshio Hase's neighbourhood. Together the case studies cover local and national government EIA systems, pre- and post-directive EIAs, and a wide range of interest group responses to development.

With the exception of the Kansai airport, the case studies all concern inadequate EIAs. We did not set out to select the worst cases. In fact, in meetings with the EA we asked them to name the best EIA that they were aware of, or projects that had been cancelled as a result of EIA, but they were unable to do so.[7]

Table 9.2 lists the project details, estimated cost, project undertaker and EIA procedures used for the case studies. The Honshu–Shikoku Bridges, Kansai International Airport and Trans-Tokyo Bay Highway are all very large projects, with estimated costs of $24 billion, $8.5 billion, and $8 billion respectively. In comparison, the Kyoto circular highway and New Ishigaki Airport are much smaller, at $714 million and $23 million respectively. The first three projects are also large and technically complex, the bridges due to their length, and the KIA and TBH due to construction difficulties.

The assessment of the Honshu–Shikoku Bridges was completed in 1978. The EIA was based on guidelines drawn up by the EA, MoT and MoC specifically for the bridge, and marks the high point of the EA's involvement with EIA. The EA's stringent review of the EIS prompted a large number of project changes. This caused other ministries to become increasingly wary of the EA's involvement in EIA, and to severely curtail the EA's reviewing powers in the 1984 Cabinet decision.

The twenty-year planning period for the Kansai International Airport spans the range from the environmental activism of the late 1960s to the pro-development orientation of the 1980s. During these twenty years, numerous issues and interest groups rose and fell; national and local EIA procedures were developed; and three EISs were prepared for the airport between 1974 and 1986 using three sets of EIA guidelines.

The proposed New Ishigaki Airport exemplifies a smaller-scale project which would usually be of only local significance. However, the proposed airport has caused international furore because it is located in close proximity to a particularly healthy coral reef. The assessment of the NIA has not been approved to date. This case study discusses how scientific data and expertise are used in the EIA process.

**Table 9.2** Case study details

| | Honshu–Shikoku Bridge Kojima–Sakaide Route | Kansai International Airport | New Ishigaki Airport | Trans-Tokyo Bay Highway | Kyoto Second Outer Circular |
|---|---|---|---|---|---|
| Project details | 6.678km bridge | 511ha airport 3500m runway | 110ha airport 2000m runway | 15.1km bridge and tunnel | 15.7km highway |
| Estimated cost | ¥1,100 billion | ¥1,200 billion | ¥38 billion | ¥1,150 billion | ¥100 billion |
| Project undertaker | Honshu–Shikoku Bridge Authority | Kansai International Airport Company | Okinawa Prefecture | Trans-Tokyo Bay Highway Company | Kyoto Prefecture MOC |
| Relevant EIA procedures | Guidelines produced by EA, MoC, MoT | Osaka prefectural EIA guideline | PWAR Law Okinawa prefectural guidelines | MoC guidelines Kawasaki City and Kanagawa Prefectural EIA ordinances, Chiba Prefectural EIA guidelines | MoC city planning EIA guidelines |

The Trans-Tokyo Bay Highway was potentially subject to not only the 1985 MoC procedures for EIA, but also Tokyo's, Kawasaki's and Kanagawa prefectures' EIA ordinances and Chiba prefecture's guidelines. An analysis of how these systems were coordinated and applied shows that the EIS was reviewed much more stringently by the local authorities than by the national agencies, and that the local authorities were instrumental in forcing the developers to implement pollution control measures.

The assessment of the Kyoto Second Outer Circular Route was very brief, and was approved in August 1989. This example explores the interaction of interest groups in the EIA process more closely.

The case studies are arranged in roughly the following order:

— history and need for the project, economic benefit;
— impacts of the project, environmental cost;
— effectiveness of the EIA process and procedures, public participation, administrative review; and
— conclusions.

### Notes

1. British Embassy, Tokyo (1988), p. 1.
2. *Japan Times* (30 May 1988).
3. *Newton* (Special edn, Feb. 1988, in Japanese), 'Japan in the 21st century: part 1', 3(8), pp. 158–9.
4. *Days Japan* (10 Nov. 1988), vol. 457, p. 3.
5. The Akashi Kaikyo Bridge will be 1,990m long. The Humber Bridge, currently the longest in the world, is 1,410m long. Honshu–Shikoku Bridge Authority (1986), p. 5.
6. In comparison, the Channel Tunnel will be 52km long, with 36km under water: Namiki (1985).
7. The officials referred to the Nakaumi project as Lake Shinyi (project to change the lake from brackish to fresh water) as an example of a project cancelled because of its adverse environmental impact. However, the project, which was abandoned in 1988, was the subject of intense local opposition on economic grounds and in reality environmental factors were not of great significance.

# Chapter ten

# Honshu–Shikoku Road/Rail Bridges

## Introduction

The Honshu–Shikoku Bridge project involves the construction of three sets of bridges at a total cost of ¥3.3 trillion. Construction of the Akashi–Naruto route began in 1976 after a simple environmental assessment was performed under the supervision of the EA.[1] Construction on the Onomichi–Imabari route began in 1977 and on the Kojima–Sakaide route in 1978. The Kojima–Sakaide route opened in 1987 and the two other routes are expected to open in 1998–9. This case study examines the EIA process for the central Kojima–Sakaide route only. The route has a total length of 37km and is expected to cost about ¥1.1 trillion. Figure 10.1 shows the three routes.

## History and siting of the bridges

Japan's Shikoku Island has been suffering from economic decline and outward migration of population for many years. When in 1955 the Japan National Railways began exploring the possibility of building a bridge linking Shikoku to prospering Honshu, local authorities welcomed the prospect of economic revitalization. Four years later, in 1959, the MoC began comparing five potential routes on the basis of cost, economic effects, and planning and construction timescales. Environmental factors were not considered despite the fact that the proposed routes were all located in the Seto Inland Sea National Park.[2]

In 1968 the MoT and MoC agreed to construct three sets of bridges simultaneously. Pressure from influential politicians played a major part in this decision. All three routes were supported by LDP politicians: former prime ministers Miki and Ohira represented Tokushima and Kagawa prefectures (the Akashi–Naruto and Kojima–Sakaide routes respectively) and another influential politician, Mr Miyazawa, represented Hiroshima prefecture (Onomichi–Imabari route).[3] The MoT and MoC were unable to reject any of the proposed sites for fear of upsetting these

**Figure 10.1** Honshu–Shikoku Bridges project

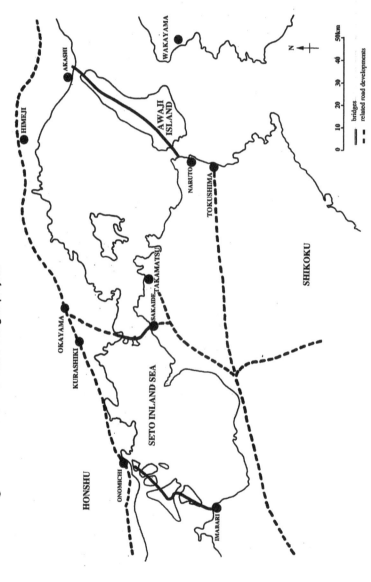

politicians. The ministries' final decision, therefore, was highly political with little concern for the environment. Opponents to the scheme argued there was only enough traffic between Honshu and Shikoku to warrant one bridge and that the siting of three sets of bridges within 200km did not justify the investment involved.

The Honshu–Shikoku Bridge Authority, a public corporation whose employees were drafted mainly from the MoT and MoC, was set up in July 1970 to supervise construction of the bridges. The authority hired several consultants to analyse the economics of the construction programme, but again environmental factors were not considered. In 1972 detailed plans for the bridges were finalized. A ceremony to mark the beginning of construction was planned for November 1973, but the scheme was shelved following the OPEC oil crisis of that year. The project returned to the political agenda in 1977, and that December construction began on the Onomichi–Imabari route.

### EIA of the Kojima–Sakaide route

At a Cabinet meeting in April 1977, the government decided informally to begin studies on the environmental effects of building the Kojima–Sakaide route. Cooperation was requested from the EA, MoT and MoC to ensure that the bridge authority received guidance on how to prepare an EIA for the bridge. The EA presented 'Basic Guidelines for Assessing the Environmental Impact of the Honshu–Shikoku Bridge Project (Kojima–Sakaide Route)' to the authority on 20 July, and the MoT and MoC provided technical guidelines and a detailed survey plan on 21 September.[4]

The EA prescribed the following procedures:

— The authority should undertake an environmental assessment and produce a draft report detailing the need for the project, scope of the assessment and proposed environmental protection measures.
— The authority should send the draft EIS to, and seek opinions from, the governors of Okayama and Kagawa prefectures, heads of local cities, towns and villages concerned, and the EA's director. In consultation with the governors, it should then open the draft EIS for public inspection, publicize its results, and consider the views of local residents.
— The authority should then draw up a final EIS, considering the opinions expressed in the consultation procedures. Copies of the published report should be sent to the local authorities and the EA's director.[5]

An interesting point about these procedures is that they did not require consultation with the competent ministry (MoT, MoC), which the later EIA guidelines require. The MoT and MoC felt that the EA usurped their traditional jurisdictional powers when assessing the bridge scheme. It has been suggested that this experience brought about a general reluctance on the part of the ministries involved to cooperate in the development of EIA procedures.

The EA specified the matters which the authority should consider in the assessment, namely:

— the present state of the environment;
— potential effects of the project on environmental quality;
— environmental protection measures to be undertaken by other bodies; and
— details of alternative plans, if any.

Table 10.1 lists environmental factors investigated in the EIS. Other factors included social and historical changes, landscape, present regulations on the natural environment, and the utility and scientific importance of the environment. Of particular interest is the need to consider alternatives, existing regulations, and the utility of the environment, none of which were included in the national EIA guidelines of 1984.

**Table 10.1**   Impacts considered in the EIA for the Kojima–Sakaide route

|  | Construction | | | Existence | | | Operation | |
|---|---|---|---|---|---|---|---|---|
|  | sea | road | rail | sea | road | rail | car | train |
| Air pollution | x | x | x |  |  |  | x |  |
| Water pollution | x |  |  | x |  |  | x |  |
| Noise | x | x | x |  |  |  | x | x |
| Vibration | x | x | x |  |  |  | x | x |
| Low-freq. vibration |  |  |  |  |  |  | x |  |
| Ground subsidence |  |  |  |  |  | x |  |  |
| Topography |  |  |  |  |  | x | x |  |
| Plants | x | x | x |  |  | x | x | x |
| Animals | x |  |  |  | x | x | x | x |
| Scenery | x |  |  |  | x | x | x |  |
| Accidents | x | x | x |  |  |  |  |  |

Source: Oshima (1989).

The bridge authority prepared a draft EIS within five months and, after establishing procedures for public inspection and hearings through consultations with the prefectural governors, the report was made public on 19 November 1977. The EIS was available for public inspection in eleven places in local authority and public corporation offices, and twelve explanatory meetings were held from 12 November to 12 December. The authority received 1925 written opinions from individuals and organizations between 22 November and 19 December.[6]

The local authorities and the EA were asked to comment on the draft EIS from 14 January to 27 March 1978.[7] The EA criticized the assessment on five points:

— Air pollution from the additional 48,000 cars/day would cause $NO_2$ levels to exceed environmental quality standards.[8]
— Noise levels and bridge vibration while trains crossed needed to be reduced.
— Economic valuations of local scenic quality should be reassessed taking into account conditions in the area without the bridge.
— The siting of proposed interchanges should be reconsidered because they were located too near residential areas. Siting should be discussed with local residents.
— A tunnel should replace an unsightly cutting in the Mt Washu Natural Park area.

The bridge authority took until 4 May 1978 to complete and submit the final report, which did not differ substantially from the draft. In the report, the authority responded to the EA's criticism with the following comments:

— It admitted that $NO_2$ levels would exceed the existing EQS of 0.02ppm, but argued that they would not exceed World Health Organization guidelines and that the EA would soon raise the EQS for $NO_2$ to 0.04–0.06ppm.
— It also admitted that noise levels would exceed those permitted for new highways but noted that measures (unspecified) would be used to reduce noise from 85 to 80 phons.
— A tunnel at Mt Washu was considered economically unacceptable. Instead the authority proposed measures to minimize damage to scenery and provide more landscaping.[9]

Under the Natural Parks Law, the natural parks authorities must be consulted when a development is proposed for a park area. Consequently, the Natural Park Protection Council reviewed the final EIS for the Kojima–Sakaide bridge proposal. Although the council could effectively have vetoed the project, it felt that this would be unrealistic for a project of such national significance to which the central administration was so committed. However, the council proposed nine conditions to reduce the bridge's effect on the scenic beauty of the area:

— tunneling should be used in scenic areas such as the Mt Washu region;
— suspension bridges should be minimized to avoid scenic disruption;
— the bridges should be painted to blend with surrounding scenery;
— night lighting should be minimized;

— construction noise should be monitored to ensure that it does not harm marine life;
— consultation with the local authorities and local residents should continue throughout construction;
— equal consideration should be given to environmental protection inside and outside the natural park areas;
— all temporary buildings should be removed after construction; and
— after construction, the natural environment should be restored to good condition.[10]

In the period leading up to September 1978, the EA met with the MoT and MoC at least ten times to discuss the above matters. Finally all three bodies agreed that the bridge authority should meet the following four conditions:

— the authority should purchase and landscape all land within 30m of the highways;
— it should build roads and parks in the Mt Washu area to enhance the natural characteristics of the area and promote tourism;
— a fund for the environmental protection of the area should be set up; and
— plans for environmental protection should be prepared by a council composed of representatives from the two prefectures, the bridge authority and the two ministries.

When the required consultations were completed and approval had been granted by the concerned ministries and local authorities, construction began on 10 October 1978. The ceremony was boycotted by local protest groups which claimed that the environmental protection measures were insufficient.[11] The bridge was completed on 10 March 1988.

### Conclusions

The Honshu–Shikoku Bridge project was a testing ground for the introduction of national EIA procedures into Japanese development planning. The EA adopted a strong stance when producing guidelines and negotiating with the bridge authority and other government organizations. However, the bridge authority, MoC and MoT opposed the EA's stance because they feared increased costs and delays and because they felt that the EA was infringing on their jurisdictions. Eventually the bridge authority agreed to the four conditions listed above; although this was an achievement in itself and costly, it did little to minimize the impact of the whole scheme.

Efforts to include the views of local residents, to consider alternatives and to perform a comprehensive assessment are admirable. Unfortunately, the EA did not have power to ensure that these measures were properly carried out. The bridge authority ignored many of the views expressed by local residents, and only partially analysed the bridge's effects on pollution (particularly noise) and on scenery. The legacy of this became clear recently when the bridge opened and nearby residents began to complain about unacceptable noise levels, especially in the early morning.[12]

## Notes

1. Environment Agency (Mar. 1987).
2. Several Japanese scientists requested that the start of construction of the Oshima Bridge be delayed until comprehensive investigations on the ecology of the inland sea were undertaken, but their requests were ignored.
3. Hase (1981b), p. 242.
4. Ibid., p. 247.
5. Oshima (1989).
6. Environment Agency (Aug. 1977).
7. Oshima (1989).
8. In reality, only 11,000 cars/day cross the bridge due to the high tariffs.
9. Environment Agency (Aug. 1978).
10. Hase (1981b), p. 246.
11. Ibid., p. 247. For example, residents in the Heida area wanted a tunnel to reduce noise and air pollution but their demands were refused. On Shikoku, the residents of Sakaide City protested because their request for more careful monitoring of possible public health effects associated with interchanges close to their communities had been ignored.
12. Noise levels in residential areas directly below the bridge range from 80 to 90 phons. The 102 daily train crossings are the main source of this noise. Local citizen protest in June 1988 led to the termination of four trains in the early morning and late evening: Nakamura (1989), p. 196.

# Chapter eleven

# Kansai International Airport

## Introduction

Construction is already under way on the first phase of the Kansai International Airport (KIA). The airport, which is scheduled for completion in 1993, will be located on a 511ha reclaimed island connected to the mainland by a 3.8km road/rail bridge. At a cost of ¥1.19 trillion ($8.5 billion), the KIA is expected to revive the economic vitality and boost the international status of the Kansai region[1] by providing Japan's only 24-hour airport. Figure 11.1 shows the KIA and associated developments, and Table 11.1 is a chronology of the KIA's development.

## Economic arguments for a new Kansai airport

In the 1960s the MoT concluded that the expansion of air transport facilities was particularly needed in the Kansai area. The number of air passengers in the area was predicted to increase from 5 million in 1968 to just over 12 million in 1985, greatly exceeding the capacity of the existing Osaka International Airport. Further expansion of the Osaka airport was unfeasible: the airport is located in a densely populated area of western Osaka (see Figure 11.2) and has been the subject of a series of lawsuits from local residents over noise and air pollution problems, especially since the introduction of jets in 1964.[2] As a consequence of these legal actions the airport's operating hours were restricted to 7am–9pm in 1976, and takeoffs and landings were limited to 200 jets/day in 1977. The noise problems associated with the Osaka airport were a major reason for planning a new airport and partly explain the desirability of constructing a sea-based airport.

Another justification for a new airport was the comparatively poor performance of the Kansai economy. The local economy is dependent on export-oriented industries, textiles, metal manufacture and shipbuilding. Both the national government and local financial circles hoped that a new airport would encourage the restructuring of the economy towards more

**Figure 11.1**   Kansai International Airport and Maejima

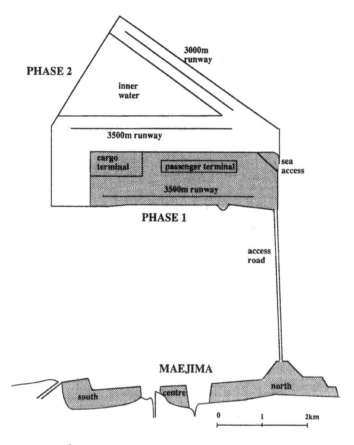

high-tech industry.[3] In the fourth Comprehensive National Development Plan of 1987, the KIA and associated developments are described as a main pillar in the development of Kansai region as an advanced economic, cultural and academic centre.

The airport will have a considerable economic impact. According to a 1983 report by Osaka prefecture, in addition to the airport cost of ¥1.19 trillion, another ¥2.25 trillion will be used to finance associated local public sector developments. These include Maejima, a 318ha reclamation on the shore of Osaka Bay adjacent to the airport, which will be developed into a new community to service the airport (see Figure 11.1). In addition, road and rail links to the airport will be provided. Sites in Osaka, Wakayama and Hyogo prefectures, from which soil will be excavated for filling material for the reclamation, will be developed as

**Table 11.1**   Chronology of KIA-related events

| Year | Month | |
|------|-------|---|
| 1965 | | Need for KIA discussed. |
| 1968 | 8 | MoT proposes eight possible sites for KIA. |
| 1971 | 10 | MoT establishes KIA division in Aviation Deliberation Council to study location and scale of KIA. |
| 1972 | 8 | MoT holds three-day public hearing concerning Kobe and Senshu sites. |
| 1974 | 8 | ADC reports to MoT: Senshu is preferred site, assuming closure of Osaka airport. |
| 1975 | 6 | MoT makes ADC's report public. |
| 1976 | 7 | Osaka governor proposes conditions for accepting monitoring towers; MoT announces environmental monitoring plan. |
| 1980 | 7 | ADC report recommends reclamation. |
| 1981 | 4 | MoT presents 'Three-Point Set' (airport plan, EIS and regional development plan) to Osaka, Hyogo and Wakayama prefs. |
| 1982 | 4 | MoT holds seven public hearings about KIA. |
| | 7–8 | Osaka and Wakayama prefs. approve 'Three-Point Set'. |
| 1984 | 2 | Hyogo pref. approves 'Three-Point Set'. |
| | 10 | KIA Co. established. |
| 1985 | 10 | KIA Co. presents draft EIS to Osaka pref. and holds five explanatory meetings. Osaka pref. releases draft EIS for Maejima. |
| | 12 | EISs for KIA and Maejima discussed at three public hearings. |
| 1986 | 2 | KIA environmental surveillance organization established. |
| | 4 | Osaka governor gives opinion on KIA. |
| | 6 | KIA Co. and Osaka pref. release final EISs for KIA and Maejima. |
| | 7 | KIA Co. applies to Osaka pref. for permission to reclaim land, and to MoT for permission to establish airport. |
| | 11 | Local authorities consent to land reclamation; Osaka pref. presents reclamation application to MoC. |
| | 12 | MoT permits KIA establishment. |
| 1987 | 1 | MoC permits land reclamation; construction begins. |
| 1993 | | Projected completion date of KIA. |

residential and park areas. It has been estimated that the economic effect of the airport on the Kansai region will be around ¥4.6 trillion/year.[4]

### Environmental arguments against the KIA

The late 1960s and early 1970s were a time of nationwide environmental activism in Japan, a time during which the traditional emphasis on economic growth was offset by growing citizen concern for the environment. In the early days of the KIA's planning, the main complaint of local citizen groups concerned the government's secrecy and lack of public consultation. The MoT selected both the site and the construction method for the airport with little public consultation. By not making its decision-making public, the MoT avoided giving the citizen movements something concrete to protest against, thus keeping them disorganized and minimizing opposition to any given plan. As the airport's planning progressed and more information became available, the protest groups expressed concern over the airport's potential social, economic and environmental effects, which included:

— the role of the KIA in Japan's transport system;
— development pressures on nearby parkland associated with increased economic activity;
— changes in the traditional lifestyle of the local population;
— noise and air pollution generated by the KIA and associated developments;
— water pollution and disturbance of the currents in Osaka Bay caused by the reclamation work, and later increases in sewerage and run-off;
— availability of the 33,000 tons/day of water needed for the airport's operation;
— damage caused by soil excavation works, and by the transport of soil to the airport site;
— danger to shipping from the airport's noise and bright lights, and the airport-related increase in sea traffic; and
— the airport's stability during major natural disasters such as earthquakes and tidal waves.[5]

### Siting of the KIA

The KIA began to be widely discussed in 1965 as a means to overcome the noise problems at the OIA.[6] The MoT began conducting surveys in 1968 and set up the Aviation Deliberation Council (ADC) in 1971. The ADC was composed of seventeen academics, industrialists, researchers and government officials. It was given the task of deciding the location and size of the new airport within the framework of four specified goals, namely:

— ability to handle international and domestic flights;
— round-the-clock air transport facilities;
— comprehensive environmental protection; and
— possibility of further expansion.

The ADC took three years to complete its deliberations.[7] Seven sites were initially considered, as shown in Figure 11.2. These were later narrowed down to three: two sites were rejected because they were too close to urban centres, another two sites were merged into one, and Awaji Island was rejected because it had a large population living within the airport's 70 WECPNL noise range. The three remaining sites were all located in the sea at Senshu, Kobe and Harimanada.

Environmental (noise, water and air pollution) and social conditions at each site were then investigated and the ADC held a series of public hearings in mid-1972 to determine local opinions. Seven criteria were

**Figure 11.2**  Alternative sites for the Kansai International Airport

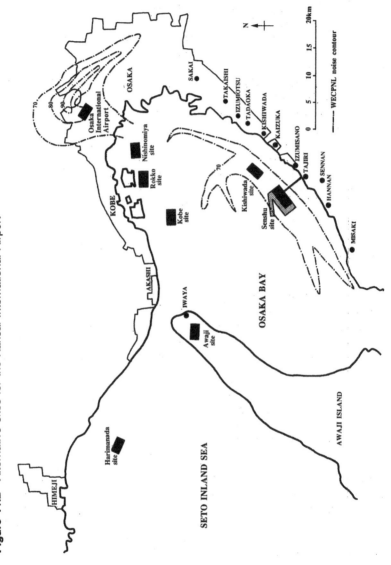

used to compare the three sites: convenience of use, air traffic control, environmental factors, construction, coordination with existing rights and interests, coordination with local plans, and the effect of development on the local economy. Another criterion, local receptivity to the airport, was excluded on the grounds that it was a politically sensitive issue and therefore outside the remit of the ADC.[8] Table 11.2 summarizes the

**Table 11.2**   Summary of ADC's comparison of candidate sites for KIA

|  | Senshu | Kobe | Harimanada |
|---|---|---|---|
| Convenience | If Osaka Airport was used by 100 people, other sites' users would be: 80 / 85 / 55. Senshu and Kobe are within range of Osaka–Kobe transport system, but not Harimanada. | | |
| Air traffic control | Sites' annual flight capacities are: 150–160,000 | 150–160,000 | 190–200,000 |
|  | Due to wind impact, runway use would be: 93.6% | 98.1% | 98.9% |
|  | Kobe site limited by Akashi bridge, Osaka city etc. | | |
| Environmental | Impact on current, water quality, and natural environment would not be marked for any site. At Kobe, airplane noise would meet environmental standards but 70WECPNL contour would come near Kobe and Awaji Island. Present $NO_2$ levels are: 0.006ppm | 0.04ppm | 0.036ppm |
|  | Kobe and Harimanada would have problems meeting EA's total $NO_2$ standards. Ventilation tower for exhaust gas of Kobe's land-airport tunnel would be a local problem. | | |
| Construction | Sea bottom weak. Bridge can be built. | Sea bottom weak. Tunnel needed due to marine traffic. | Sea bottom satis- factory. Bridge can be built. |
|  | Assuming bridge for Senshu and Harimanada and tunnel for Kobe, not including airport facilities, projected cost (1972 values) would be: ¥570 million | ¥570 million | ¥320 million |
| Coordination with local rights | Impact on marine traffic would be small. Osaka's fishing industry is small but flourishing. | Site adjoins Kobe and Osaka's marine routes and would cause great impact. Fishing rights and urbanization would eliminate a large area. | East and North Harimanada's marine routes would be affected. Excellent fishing areas; fishing industry is high-grade. |
| Coordination with local plans | In line with local plans if transport network is completed. | Coordination with plans for population and marine industry would be difficult. | Coordination of transport network difficult due to condition of bayside industrial area and distance to city. |
| Effectiveness of development | Corresponds to coordination with local plans. | | |

Source: Ministry of Transport (Aug. 1974), pp. 40, 52a.

**Table 11.3** ADC process for determining KIA site

| Criterion | a | Senshu b | c | Kobe b | c | Harimanada b | c |
|---|---|---|---|---|---|---|---|
| Convenience of use | 21.7 | 139.5 | 3027 | 152.0 | 3298 | 95.5 | 2072 |
| Traffic control | 19.9 | 137.5 | 2736 | 124.0 | 2468 | 155.0 | 3084 |
| Environment | 18.8 | 143.0 | 2688 | 119.0 | 2237 | 141.0 | 2651 |
| Construction | 12.4 | 132.9 | 1648 | 119.0 | 1476 | 145.0 | 1798 |
| Rights & interests | 8.9 | 144.9 | 1290 | 113.5 | 1010 | 105.0 | 935 |
| Local coordination | 9.2 | 147.0 | 1352 | 111.0 | 1021 | 132.5 | 1219 |
| Development effects | 1.6 | 144.5 | 1315 | 109.5 | 996 | 127.5 | 1160 |
| Total (in 1,000s) | 100.0 | | 14056 | | 12506 | | 12919 |

a = percentage breakdown of 17,000 points allocated by the ADC members.
b = number of points from total of 170 (each member could allocate up to 10 points for each criterion) which indicates how well the site fulfilled each criterion.
c = a x b

Source: Ministry of Transport (Aug. 1974), pp. 54–6.

ADC's comparison of sites.

In making its final decision, the ADC used the following procedure. First, each of the 17 ADC members allocated 1,000 points (total 17,000) between the seven criteria to determine their relative importance (see Table 11.3, column a for percentage breakdown). Second, each member judged each site on a scale of 0 to 10 to determine how well it fulfilled each criterion (column b). Finally, for each site, the importance criteria were multiplied by the fulfilment criteria (a × b).[9]

As can be seen in Table 11.3, Senshu rates only marginally better than the other two sites on environmental grounds, but scores higher in relation to existing rights (its effect on marine traffic and the marine industry were considered to be minimal), coordination with local plans and effect on the local economy. Kobe was considered more convenient to use, and Harimanada was superior in terms of air traffic control and ease of construction.

In August 1974 the ADC released its report which concluded that, presupposing the closure of the Osaka airport, a sea-based airport 5km off Senshu would be most appropriate.

**Local authority negotiations**

According to the procedures set down in the Public Water Areas Reclamation Act, a developer must apply to the local governor for permission to begin reclamation. In turn, the governor must obtain the agreement of the local mayors before approving the project. In the early 1970s, thirteen of the fifteen local authorities affected, including Osaka prefecture, were opposed to the KIA proposal.[10] By 1982, all of them (except Hyogo prefecture, which took until Feb. 1984) had agreed to the scheme.

Osaka prefecture's response to the ADC report of 1974 was rather neutral: the ADC was merely requested to release the material on which its decision was based. When the ADC complied in June 1975, the prefecture and the EA criticized the lack of environmental data and requested further studies. A period of negotiation between the local authorities and the MoT then took place in which the prefecture forcefully bargained for measures which would help mitigate the airport's potential effects. As a result the MoT agreed in 1976 to

— install towers for environmental monitoring;
— undertake comprehensive studies of local conditions, including social and economic factors;
— establish a deliberation body with citizen participation; and
— set up procedures for local authority verification of the above activities.[11]

Thus the local authorities were able to force the MoT to consider the wider effects of the project, as well as to undergo a more stringent review process. In addition, several prefectural advisory committees were set up to report on such issues as traffic forecasts, fishing, air and water quality, and infrastructure plans.[12]

Two factors were, however, gradually undermining the opposition of the local authorities. First, the local economy was gradually deteriorating. The nationwide economic slump caused by the oil shock of 1973/4 was accentuated in Kansai where Osaka prefecture had been operating at a deficit in order to implement its welfare policies. Second, well-publicized decreases in certain pollutants caused the public to believe that pollution problems had been solved. Thus environmental concerns lost much of their importance while economic concerns grew. By the late 1970s Kansai's political climate had become more conservative and in the 1979 gubernatorial elections Osaka's activist governor, Sakae Kishi, was defeated. Leaders of the smaller local authorities quickly realized the futility of opposing the airport plan.

However, the fact that the project required local authority approval meant that the authorities were in a strong position to bargain with the MoT for financial assistance for local infrastructure projects. The negotiations which took place between the local authorities, national government and industry during this time were characteristically Japanese in style. Such negotiations normally involve a large number of meetings between the parties involved, with the stronger party gradually wearing down the opposition of the weaker party. The stronger party's increasing social and political pressure, its willingness to offer minor financial concessions, and most effectively its willingness to draw out the length of the negotiation period, all work to wear down opposition. Because local authorities receive much of their funding from the national government and because the national government has strong links with industry, negotiations between these parties often boil down to a discussion of how much compensation the local authorities should receive.

By 1985 Osaka prefecture's plans for the area included extensions to the rail network, new railway stations, parks, a dam, a new town serving the airport and various housing developments. All of these projects are subsidized by the national government. In addition, the authorities nearest the airport received ¥1.5 billion for the maintenance of the local infrastructure.[13]

### EIA of the KIA

The first official EIA for the airport, prepared according to the 1974 interim national EIA guidelines,[14] was completed by the MoT in April 1981. It was based on the findings of a second report by the ADC commissioned by the MoT in late 1979, and on studies requested by the local authorities. The second ADC report was completed in 1980 and considered runway length, direction and capacity, flight routes, construction techniques, and airport design and layout. Reclamation was recommended over a previously untried 'floating island method' as the best construction technique. Although more environmentally damaging, reclamation was considered more cost-effective and had already been successfully used in constructing the Nagasaki Airport.[15]

The EIA was 300 pages long and, although it included more environmental information than previous reports, it essentially reiterated the same points. It considered only the Senshu site and the airport size recommended by the ADC. Its main conclusion was that an airport located at Senshu would neither significantly affect the daily life of nearby residents nor hinder the attainment and maintenance of environmental standards.[16] In 1981–2 the local authorities inspected the report and in July 1982 the basic plan was approved.

In 1984 the KIA Co. Ltd was established as Japan's first private share-issuing company to construct, own and manage an airport. The company is jointly financed by the national and local governments and local companies, and its staff is 'borrowed' from various ministries and private companies. This scheme, it is suggested, combines governmental expertise with the efficiency of the private management.[17]

The final plan for Phase 1 of the airport calls for a 511ha reclaimed island with a 3,500m runway, terminals and associated buildings. The capacity of the airport will be approximately 160,000 take-offs and landings annually. Phase 2 would expand the site to 1,200ha, with two additional runways of 3,500m and 3,000m, as shown in Figure 11.1.[18]

The KIA Co. needed to obtain consent from the MoT and the EA for the airport operation and construction; and from Osaka prefecture and the MoC for reclamation. The Public Water Areas Reclamation Act requires that the developer produce an EIA for the proposed development. This EIA is then included with the application for reclamation which the developer submits to the prefecture. When preparing the second official EIA for the airport, the KIA Co. used Osaka's 1984 guidelines, shown in Figure 11.3. These guidelines require a project developer to make a draft EIS public for one month and hold explanatory meetings. In addition, for a period of six weeks after the report's release, residents of the concerned area can send their written comments to the developer, and citizens (not necessarily local residents) can send their opinions to the governor. The governor then confers with the prefectural environment committee and local mayors before deciding whether or not to approve the plan.[19]

The draft EIS for the KIA was released in October 1985. It was very similar, in both format and content, to the MoT's EIS of 1981: the range of environmental factors considered was slightly expanded, the impact of the access bridge was discussed, and airplane noise was re-estimated to consider recent improvements in aircraft design and new assumptions on wind direction and flight altitude. The EIS considered only Phase 1 of the development.

Five explanatory meetings were held during October 1985. The company also explained the draft EIS at approximately thirty local government and community group meetings. Throughout October and November, the company and Osaka prefecture received citizens' written opinions. To gain more public opinion, Osaka prefecture and the other local authorities also held three public hearings on the KIA issue in December 1985 and January 1986. In April 1986 Osaka's governor made recommendations on the draft EIS, and in June the KIA Co. released the final – approved – EIS. The MoC approved the project in late 1986, and Osaka prefecture gave the KIA Co. permission to begin construction in January 1987.

EIAs were also prepared for other airport-related developments. Osaka prefecture released plans for Maejima in June 1985, produced a draft EIS in October, and released the final EIS in June 1986. Osaka and Wakayama prefectures produced EISs for the soil excavation areas and subsequent developments in their jurisdictions.

**Effectiveness of the EIA procedures**

*Timing and consideration of alternatives*

The ADC, on behalf of the MoT, prepared two reports which provided background data, criteria for choice, and an explanation of the reasons for selecting the Senshu site and the airport size. The timing of the

**Figure 11.3**   Osaka prefecture EIA procedures

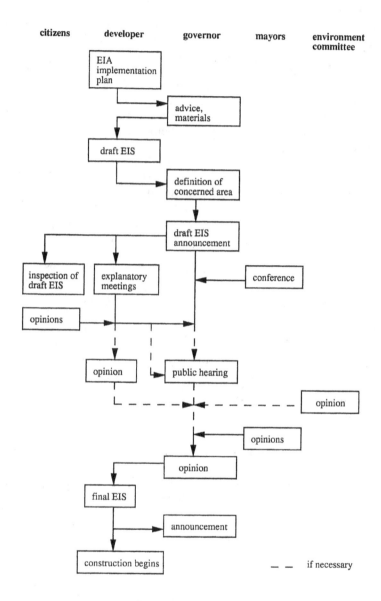

Source: Osaka Prefectural Government (April 1984).

ADC's reports was correct, and the ADC can be commended on having accomplished (albeit crudely) the difficult task of quantifying the importance of environmental factors and incorporating them in the decision-making process. The fact that the final decision to site the airport at Senshu was based primarily on the opinions of seventeen knowledge-able but possibly biased men with limited public consultation does, however, raise some doubts about the form of democratic decision-making in Japan.

The two later assessments, by the MoT in 1981 and the KIA Co. in 1985, go much further in their consideration of environmental factors than the ADC's reports. However, both EIAs are based on the assumption that an airport should be built at Senshu on reclaimed land. They do not reconsider the factor which has the greatest environmental influence, namely location. They serve only to justify the original proposal and, at best, suggest measures to mitigate the environmental impact of that decision.

*Comprehensiveness*

The range of environmental impacts considered in the 1986 EIS is shown in Table 11.4. The EIS lacks comprehensiveness for several reasons:

— Only the impact of the airport and the connecting bridge were considered. Other related developments are covered by separate EIAs, undertaken by different bodies; secondary and synergistic impacts are ignored.
— The 1986 EIS discusses only Phase 1 of the airport without considering possible future expansion. Phase 2 was discussed in the 1981 EIS, which showed that the larger airport would have a greater environmental impact. Perhaps the reason for the later EIS's omission of Phase 2 can be partly explained with reference to $NO_2$ predictions. According to Osaka prefecture's environmental standards, the max-imum $NO_2$ release for any given development must not exceed 7,640 tons/year. Phase 1 of the KIA (511ha, one runway) was estimated to produce 4,350 tons/year. If the addition of Phase 2 (689ha, two runways) merely doubled the airport's $NO_2$ emissions, it would cause emissions to exceed 7,640 tons/year. The question is whether Phase 2 would then be refused or whether the environmental standard would be modified.[20]

*Method of assessment*

The MoT and the KIA spent a great deal of money (¥8 billion and ¥10 billion respectively) to prepare the EIAs of 1981 and 1986. Much reliance was placed on complicated computer simulations. However, these

**Table 11.4**   Environmental impacts considered in the 1986 EIS for the KIA

*Construction (reclamation and access bridge):*
— noise
— air quality ($SO_2$, $NO_2$, suspended particulates)
— water quality (turbidity, dumping of earth/sand)
— marine life
— shipping

*Existence (airport and access bridge) and activities (airline service, use of airport facilities, sea access by tankers, bridge traffic)*
— noise (aircraft, road traffic, railway, low frequency)
— air quality ($SO_2$, $NO_2$, CO, suspended particulates, photochemical oxidants, hydrocarbons)
— currents
— water quality (COD, salinity, pH etc., hazardous substances)
— marine life
— marine phenomena (waves, high tide, tsunami, configuration of seashore)
— land animals (birds)
— scenery
— electricity
— shipping

Source: Kansai International Airport Co. (June 1986), pp. 1–6.

simulations depend on assumptions which may not be valid. An example is the predictions of the KIA's noise levels, as shown in Figure 11.2. At the public hearings the KIA Co. maintained that noise levels on land would be at most 60–70 WECPNL. However, flight tests by the Civil Aviation Bureau showed that if departing jets veered slightly from the correct flight course, noise on land could exceed 70 WECPNL.[21] The computer simulations did not take into account these potential changes in flight course.

*Public participation*

The level of public participation varied throughout the project's history. The 1974 report was prepared with almost no public involvement. The ADC held only one two-day meeting, mainly to determine the relative merits of the Osaka and Kobe sites. Public 'participation' at that time, lacking a constructive channel, took the form of protests and rallies.[22]

The later EIAs had broader public participation procedures. Between May 1981 and July 1982 Osaka prefecture held more than 100 meetings with local town councils, committees, and farming, fishing and economic groups. The town councils, in turn, held public meetings. The MoT held seven public hearings in April 1982 at which residents could express their views. However, the majority of the chosen speakers were pro-airport while the majority of the audience were anti-airport. Most importantly, only one-fifth of those who wanted to attend were allowed to do so. The protesters complained of the MoT 'black listing' from hearings those people who had earlier voiced opposition to the plan.[23]

For the 1986 EIA, the KIA Co. held five public meetings to explain the impact statement. The public was also given the opportunity to express its views in written form to the company and Osaka's governor, and again at meetings held by Osaka prefecture in December 1985 and January 1986. The extent to which this participation had any influence on the decision-making process for the airport is questionable. The meetings seemed primarily directed at convincing the residents of the need for the airport and its 'insignificant' environmental impact, which will be compensated for anyway by the predicted boost to the local economy. The MoT also later stated that the high demand for the Osaka airport calls for its retention despite local opposition.[24] Such an about-face undermines the original justification for the KIA and sheds doubt on the whole decision-making procedure for the airport.

Radical groups took a less constructive view of 'public participation'. For example in April 1984 radicals were blamed for two explosions in KIA-related offices, and in August 1987 they were held responsible for bombing a KIA Co. ship. This prompted the government to guard the start of the KIA's construction with 700 police, forty ships and five helicopters.

*Administrative review*

As discussed earlier, negotiations between the local authorities and the MoT/KIA Co. were invaluable in forcing environmental factors to be considered in the decision-making process. However, given Japan's political and economic situation, the local authorities had little choice but to approve the airport while trying to extract the maximum benefit for their areas.

The EA must be consulted for any reclamation scheme over 50ha. Although the EA did request additional data after reviewing the 1974 report, it neither used its veto power nor attempted to make any major modifications to the airport plan. The MoT, as the ministry responsible for the airport's siting, cannot be expected to review the assessment without bias.

**Conclusions**

No fully comprehensive EIA, based on adequate data and properly reviewed, has been prepared for the KIA, although the sum of the three reports comes close. If the alternative sites considered in the 1974 report had been compared based on environmental data such as that in the 1986 EIA, if the resulting EIS had been scrutinized by the local authorities as the 1981 EIS was, and if citizen participation was expanded, then the airport's EIA would have been a model of comprehensive environmental planning. As it is, one can only remain frustrated at the unfulfilled potential of Japan's EIA system.

The Japanese government has attempted to construct a framework of democratic decision-making with regard to development policy, but in reality this framework is often undermined by the government's autocratic nature. The KIA can be considered, in some respects, an example of good planning. Unfortunately, the negotiation and decision-making procedure leading up to the project's approval characterizes the darker side of development policy in Japan. Although the KIA's environmental consequences were acceptable to the developer and the decision-maker, they may not have been to most local residents. The local authorities, though initially against the proposal, soon became pro-development when they were offered sufficient incentives. Japan's EIA procedures seem to do little to weaken the power of the government–industry complex. Similar problems have been encountered with other airport developments, as will be shown in the case study of the New Ishigaki Airport in Chapter 12.

### Notes

1. The Kansai region is the second most densely populated area of Japan after the Tokyo (Kanto) region.
2. A series of lawsuits relating to noise problems at the airport resulted in the restriction of operating hours to 7am–10pm in 1972, forced the MoT to pay indemnity to severely affected residents in 1974, and further restricted operating hours to 7am–9pm in 1976. The Supreme Court reversed the lower court ruling on the operating hours, restoring the 7am–10pm period, in 1981: Takada (1978), pp. 35–42; *Mainichi Daily Times* (17 Dec. 1981).
3. *Mainichi Daily Times* (28 Oct. 1971, 11 Nov. 1971, 1 Oct. 1974).
4. Osaka Prefectural Government (1983).
5. *Mainichi Daily Times* (8 May 1971, 17 July 1971, 8 Nov. 1971, 28 Dec. 1971, 1 Oct. 1972).
6. Kansai International Airport Co. (May 1987).
7. Hiraiwa (1979), pp. 106–19.
8. Ibid.
9. Ministry of Transport (Aug. 1974).
10. Osaka Prefectural Government (1986b), p. 109.
11. Hiraiwa (1979), pp. 137–50.
12. Osaka Prefectural Government (1983), pp. 50–9.
13. *Asahi News* (15 Dec. 1985, 29 Nov. 1986).
14. Environment Agency (Sep. 1974), pp. 1–4.
15. Two other methods – dredging (where the airport is below sea level, surrounded by high walls) and bridge-style – were rejected earlier on financial grounds: Hiraiwa (1979), pp. 178–82.
16. Ministry of Transport (1981a), p. 269.
17. Kansai International Airport Co. (May 1987), pp. 6–8.

18. Kansai International Airport Co. (Feb. 1987).
19. Osaka Prefectural Government (Apr. 1984).
20. *Asahi News* (20 Nov. 1985, 3 Dec. 1985). Data are taken from the EIAs of 1981 and 1986, and from Osaka prefecture's environmental standards.
21. *Mainichi Daily Times* (17 Aug. 1971, 8 Nov. 1971).
22. Hiraiwa (1979), p. 66.
23. Osaka Prefectural Government (1983), pp. 110–17.
24. According to their estimates, demand for the Osaka International Airport will increase from 13 million air passengers annually in 1983 to 17 million by the year 2000: *Mainichi News* (10 Oct. 1985).

# Chapter twelve

# New Ishigaki Airport

## Introduction

Ishigaki (pop. 46,000 in 1989) is the second largest island in the nineteen-island Yaeyama group in the Ryukyu archipelago, which is located about 1,700km south-west of Tokyo (see Figure 12.1). Ishigaki has a tropical climate, and its economy is largely based on agriculture, construction and tourism. The Okinawa prefectural government plans to build a new airport with a 2,000m runway on a 100ha site. This airport will replace the existing airport and its 1,500m runway which is considered too short for jet traffic. Most of Ishigaki's residents view the proposed New Ishigaki Airport as a way to bring more tourists to the island, provide greater access for locally produced goods to mainland markets and boost the island's construction industry.

A site near Shiraho village (Shiraho marine site) had been discussed since the mid-1970s as the location for the NIA. However, this proposal had caused local – and later international – furore because it was likely to cause extensive damage to what may be the best preserved coral reef in the archipelago. The reef holds the world's largest known colony of blue coral, and Shiraho's villagers rely on it for food, building materials and the economic vitality of their village. Opposition to the scheme by the Environment Agency led Okinawa prefecture to reduce the proposed 2,500m runway by 500m in August 1987, and to move the proposed site about 3km northward (East Karadake site) in April 1989. [1] At the time of writing, the eventual outcome of the NIA is still unclear.

To date, four EIAs have been prepared for the Shiraho site, and a new EIA is being prepared for the East Karadake site. The first two were undertaken by private consultants in 1981 and 1983, and were both confidential. The third was prepared by the New Japan Marine Climate Corporation (which also prepared the second EIS) and was released in September 1986. The fourth was prepared as a result of the August 1987 changes, and was released in April 1988. Table 12.1 is a chronology of events related to the NIA's EIA.

**Figure 12.1** Alternative sites for the New Ishigaki Airport

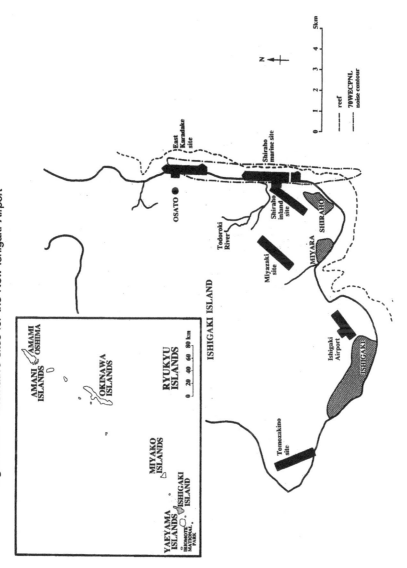

**Table 12.1**   Chronology of NIA-related events

| Year | Month | |
|------|-------|---|
| 1975 | | Ishigaki mayor mentions possibility of a 2,500m airport on reclaimed land at Shiraho. |
| 1976 | 11 | NIA plan presented to Ishigaki Town Council. |
| 1979 | 7 | New Airport Construction Promotion Group forms, visits MoF to request funds for initial surveys. |
| | 12 | Association of Shiraho United in Opposition to the New Airport forms, calls for abandonment of plan. |
| 1980 | 5 | Yaeyama Fishing Cooperative asks local authorities to change airport site. |
| | 6 | Cooperative votes to accept ¥500 million ($2 million) in compensation for the loss of fishing rights. |
| 1981 | 3 | First EIS completed. |
| | 11 | MoF approves NIA appropriations. |
| 1982 | 3 | MoT approves airport plan. |
| 1983 | | Second EIS completed. |
| | 7 | Committee to Consider the NIA Problem set up. |
| 1984 | 4 | Hearing for 33 Shiraho fishermen who sued cooperative claiming that decision to relinquish fishing rights was void because the 179 members who voted in favour didn't constitute the required two-thirds of total 945 membership. |
| | 8–9 | Prefecture tries to undertake environmental survey, but is stopped by local protest. Survey undertaken under heavy riot police protection. |
| | 10 | Cousteau Society surveys reef, publishes report criticizing NIA plan and urging further research. |
| | 11 | 'Basic Plan for the Marine Industry', commissioned from Shoei Shirai by local authorities, suggests that Shiraho coral is of low diversity. Environmental groups criticize validity of findings. |
| 1985 | 6 | Fishing cooperative votes again, by a majority, to waive fishing rights. |
| | 7 | Prefecture's engineering committee votes to promote the airport. |
| 1986 | 2 | Okinawa District Court rejects Shiraho fishermen's lawsuit. New lawsuit initiated. |
| | 3 | Okinawa pref. establishes Informal Advisory Committee to consider NIA plan. |
| | | Local mayoral elections. Shiraho candidate gains 1,101 votes. |
| | 7 | Third draft EIS available for public inspection. Informal Advisory Committee recommends that Shiraho is a suitable airport site. |
| | 8 | Explanatory meeting for the EIS held. |
| 1987 | 1 | Protesters prevent Okinawa Development Agency officials from undertaking supplementary surveys to correct errors in EIS. Riot police break up demonstration. |
| | 5 | 1,000 farmers attend rally to promote NIA. |
| | 7 | EA officials state in Diet that preservation of blue coral at Shiraho would be difficult under existing plan. |
| | 8 | Prefecture decides to reduce southern end of proposed NIA runway by 500m, but leaves open the possibility of extending it north. EA finds this sufficient to protect blue coral. Change necessitates new EIA. |
| 1988 | 2 | International Union for Nature Conservation passes resolution calling for reconsideration of NIA and designation of Shiraho reef as a nature preserve. |
| | 4 | Fourth EIS available for public inspection and comments. |
| | 5 | Explanatory meeting held at Shiraho. |
| | 6 | Local authorities receive 3,700 opinions on draft EIS. |
| | 11–12 | EA surveys state of coral. |
| 1989 | 3 | Osaka lawyers produce report on illegalities of the NIA plan and procedures. |

| Year | Month | |
|------|-------|---|
| 1989 | 4 | EA criticizes Shiraho NIA plan. Prefecture announces new site. Fifth EIA started. |
| | 6 | Locals blockade prefecture's attempts to carry out environmental surveys. |
| | 8 | Local resistance to surveys collapses. |
| | 8–11 | WWFN and IUCN survey new site. |
| 1990 | 6 | Socialist Party announces its opposition to NIA plan; recommends expansion of existing airport. |

The NIA is one of the most controversial development projects in Japan, but the issues it represents are not unique. Both the proponents and opponents of the scheme base their claims on 'objective' scientific evidence, but in the absence of an unbiased reviewer this evidence becomes a mere decorative covering for a basic conflict of interests. The effectiveness of EIA is limited by the political climate in which it is used, and the NIA's opponents face a losing battle against the ruling pro-development administration despite the strength of their scientific arguments. The NIA is an example of the continuing difficulty of preserving sensitive and unique ecosystems in the face of an administration intent on promoting short-term economic goals.

This chapter reviews the EIA process for the New Ishigaki Airport, with particular emphasis on the prefecture's EIA of April 1988. The NIA raises several additional issues:

— Shiraho's thirty-three fishermen have sued their fishing cooperative for voting to accept ¥500 million in compensation for the loss of fishing rights at Shiraho without a quorum and without undertaking a survey to assess the value of these rights.
— The MoT declined to approve the NIA plan in December 1980 because it maintained that local consensus had not been reached. However, in 1982 mayor Uchihara assured the ministry that consensus would soon be reached. The MoT approved the plan in March 1982. Since then the authorities have made no further efforts have obtain a consensus.
— Some conservationists link the government's support for the NIA with the possibility that the existing airport could be taken over by the Japanese Self-Defence Force.

For further discussion of these topics see the reports listed below.

### Economic arguments for the NIA

The NIA plan is part of Okinawa prefecture's comprehensive development strategy. The existing Ishigaki Airport serves the whole of the Yaeyamas; two smaller airports link to it. The local authorities aim to unify the Yaeyamas by providing better transport and communication links.

According to the prefecture's EISs, the number of passengers using the existing airport has steadily increased from about 230,000 in 1973 to 730,000 in 1986. The prefecture predicts, based on MoT forecasts, that by the year 2000 the demand for the airport will be about 1.65 million passengers.[3] It argues that a jet service is needed to meet this demand. It also claims that a new airport would help to expand the local agriculture, construction and tourism industries, and that it would alleviate current noise problems.[4] Indeed, at first glance the airport seems like the perfect solution to the island's problems.

However, a group of Japanese academics, the Committee to Consider the New Ishigaki Airport Problem, criticized these arguments.[5] They noted that the MoT forecasts of 1979 were straight-line extrapolations of the 1972-7 rate of passenger increase. These forecasts predicted a demand of 896,000 in 1985, 1.6 million in 1990 and 3 million in 2000. However, as shown in Figure 12.2, the actual demand in 1985 turned out to be almost 20 per cent lower than predicted. The MoT revised its forecast in 1984 and again in 1985, presenting several scenarios which were all about 30 per cent less than the 1979 predictions. This exercise demonstrates the difficulty and lack of reliability of predicting passenger demand. Air freight demand predictions were also based on straight-line extrapolations, and on available space rather than actual production levels.

**Figure 12.2** NIA actual air travel demand and MoT forecasts (10,000s)

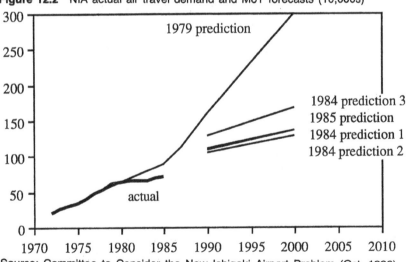

Source: Committee to Consider the New Ishigaki Airport Problem (Oct. 1986), p.17.

### Environmental arguments against the NIA

The quality of the coral reefs in the Ryukyu archipelago has gradually declined over the last twenty years, and at present most of the reefs are dead or dying.[6] The cause of this decline is a combination of Crown of Thorns starfish infestation, and sedimentation caused by run-off from development and agriculture. These problems are compounded by increased pollution from industry, agriculture, and residential development.[7] Even the marine parks have been affected.[8]

The Shiraho reef is, however, in excellent condition. It contains the world's largest and oldest known colonies of blue coral: radiographic growth studies show that the 3m-high specimens are over 600 years old. It also contains at least forty-three genera and over 100 species of hard corals (60 per cent of the coral genera found in the 2,000km-long Australian barrier reef), and more than 300 fish species, including the newly discovered blue coral goby. A species of hermit crab found there in 1987 is a designated national treasure.[9]

This good condition is due in part to the following factors:

— The reef is subject to little siltation, due to the fact that only the relatively small Todoroki River flows into the lagoon. The river water is filtered by vegetation-lined sand banks at the beach.
— The beach is almost entirely composed of bits of dead coral and shells which work as a filter.
— A greenbelt of marine vegetation in the water shallows aids filtration and provides a nursery for fish and other marine animals.
— The outer reef wall is unbroken by large channels for 12km. This may have helped to discourage starfish infestation.
— The existence of rocky fords interspersed with pools may encourage the growth of micro-atolls, of which the blue coral stands are the best known.[10]

Surveys of the East Karadake site show that the new site and the original site are part of the same ecosystem. Construction of an airport at the new site would bring about changes in tidal patterns, influencing temperature and salinity; increased turbidity, depriving the corals of light and nutrients; and an increase of starfish, algae and other coral competitors. These changes are likely to filter down to the original site and consequently the whole reef is endangered.[11]

The scheme's advocates argue that the coral will be unaffected by the development, but at a similar development on Amami Oshima all coral within 2–3km of the airport died.[12] Past efforts to use artificial filters to replicate natural processes have proven to be unsuccessful because they are too simplistic, as witnessed by the widespread destruction of other reefs in the archipelago.

### Siting of the NIA

The Shiraho site was first described as suitable for an airport by survey teams of the US Air Force before the islands reverted to Japanese control in 1968. In August 1972 Ishigaki's local authorities requested permission from Okinawa prefecture to extend the runway of the existing Ishigaki Airport to provide for a jet service, and in February 1973 jets were allowed to land there. In March 1974 a confidential study on a new airport, commissioned by the local authorities and carried out by an Okinawa consultant, was completed. This report identified five airport sites: two involved extension or readjustment of the existing airport, and the other three were at Shiraho. Although the report did not recommend any particular site, in early 1975 the mayor was already discussing the possibility of an airport with a 2,500m runway on reclaimed land at Shiraho.[13]

In November 1976 Okinawa prefecture presented three possible airport sites to Ishigaki's town council, of which the Tomizakino site was recommended as most suitable. In March 1979 the prefecture produced a report on the basic design of the airport, which considered the same three sites, but this time recommended the Shiraho site.

In July 1979 the New Airport Construction Promotion Group, composed of supporters of the scheme, decided to site the airport at Shiraho. This decision was not subject to public consultation and was criticized by local residents. In response, the prefecture carried out a brief environmental investigation in January 1980 which concluded that the airport's noise levels would not exceed environmental quality standards in Shiraho. The airport's basic plan was completed in March 1981 and was approved by the MoT a year later.[14]

The procedure by which the Shiraho site was selected is unclear because most of the local authorities' reports were confidential, and because little public consultation took place. A 1984 report by the Okinawa Development Agency retrospectively discussed how five sites – Shiraho marine, Shiraho inland, Ishigaki Airport, Miyara and Tomizakino – were evaluated. Problems with the four alternative sites are shown in Table 12.2. The report noted that these problems (e.g. air traffic control, wind direction, air route alignment) led the local authorities to choose the Shiraho marine site. However, the report did not mention any problems with the Shiraho site, perhaps because the selection procedure did not consider environmental factors other than noise.

Critics of the siting point out several inconsistencies. For example, the Tomizakino site was rejected partly because it would ruin the scenery of a historic park, but no reason was given as to why this scenery had greater value than the reef. Extension of the present airport was considered unacceptable because noise problems would make it difficult to obtain

**Table 12.2** Problems with the alternative sites for the NIA

| Sites | Noise pollution problem | Disaster prevention problem | Destruction of farmland areas | Landing lights confused with the port | Interference with national road | Plan coordination problem | | |
|---|---|---|---|---|---|---|---|---|
| | | | | | | Irrigation/drainage plans | Land improvement plans | Cultural assets protection plans |
| Present airport | x | | | | | | | |
| Tomizakino | | x | x | x | | | x | x |
| Miyara | x | x | | | | x | x | x |
| Shiraho Inland | x | | | | x | x | x | |

Source: Committee to Consider the New Ishigaki Airport Problem (October 1986), pp. 22–5.

local agreement, but the Shiraho site could be rejected on the same grounds. The other sites were unacceptable because development would interfere with agricultural land improvement schemes, but the amount of cultivated land on Ishigaki decreased by about 16 per cent between 1983 and 1987.[15]

In January 1987 the prefecture had an informal discussion with the EA, in which the EA adopted an unexpectedly strong stance and recommended that the prefecture redo its environmental studies to correct several errors. At a Diet session eight months later an EA official stated that it would be difficult to protect the blue coral colonies under the proposed plan, and asked the prefecture to alter the plan.[16] In response to this, to public pressure and to the realization that the MoT had overestimated the demand for the NIA, the prefecture decided in August 1987 to shorten the planned runway by 500m. The airport's siting was, however, not reconsidered, despite the fact that the new shorter runway would only require a 500m extension to the existing airport's runway.[17]

The prefecture planned to begin construction in February 1989. However, these plans were delayed by the EA, which decided to undertake an unprecedented two-month survey on marine conditions around Ishigaki, prompted by international pressure and news of siltation at Amami Oshima Airport.[18] In April 1989 the EA disclosed its survey results, and stated that the NIA plan could have a detrimental effect on Shiraho's coral reef. Almost simultaneously, Okinawa prefecture announced the cancellation of the Shiraho site as the preferred site for the NIA and proposed a new site at East Karadake, adjacent to Osato village (pop. 120).

The new plan effectively removes the EA from the decision-making process for the NIA. Only about 40ha of the new site will involve reclamation, and under the PWAR Law reclamation of less than 50ha does not require the MoC to consult the EA. Environmentalists still oppose the new site because they claim that its close proximity to the reef threatens the whole ecosystem.

**EIA of the NIA**

Four impact assessments were produced for the Shiraho site. The first was undertaken by Pacific Consultants and completed in March 1981. The second, carried out by the New Japan Marine Climate Corporation in 1983, quoted many of the findings of the first assessment. Neither was publicly available. The NJMCC have been involved in producing all of the EISs for the NIA since 1983. The third EIS was published in draft form by Okinawa prefecture's engineering department in September 1986. The prefecture's decision to shorten the runway required a new

EIA which was completed in April 1988.[19]

Table 12.3 briefly compares the four EIAs. The first EIA was apparently the most objective. It admitted that construction would result in a high density of suspended solids which would harm the corals and other marine life. It proposed counter-measures such as collecting ponds, chemical precipitating agents and silt-nets, but admitted that these measures were not completely effective.[20] The second EIA added some predictions on coral and water quality, but lacked a clear analysis and references, and simply stated that the proposed countermeasures would eliminate silting and thus ensure the maintenance of environmental quality standards.[21] Because the first two reports are confidential, their administrative basis remains unclear. Okinawa prefecture produced guidelines in July 1977, but their contents are not publicly available. Neither of the early EIAs included public consultation procedures, and neither can be considered comprehensive.

The third EIA of 1986 was based on the 1984 Cabinet decision and on MoC guidelines of April 1985 concerning reclamation of public waters. According to these guidelines, the prefecture must submit an EIS to the MoC with any application to reclaim land, and the MoC must consult with the MoT (and with the EA for reclamation of $\geq$50ha) before deciding whether to permit construction.[22] The 1986 EIS was available for public inspection in local authority offices from 25 July to 25 August. No copies could be made. One explanatory meeting took place on 9 August, but the hall was too small to accommodate all those who wished to attend. In total, the local authorities received 719 written comments on the EIS, of which thirteen favoured and 706 opposed the development.[23]

The fourth EIS, for the shorter runway, was essentially the same as the third. It was based on MoC technical guidelines for EIA of April 1986. The most notable change was the redefinition of pollution control measures to be used, namely 'the most recent scientific knowledge' and 'best available technology'.[24] The EIS was available for public inspection from 27 April to 13 June 1988. In total 3,699 written comments were received, of which 194 were in favour and 3,505 were opposed to the plan.[25] However, the MoC allowed only Ishigaki residents to comment,[26] and only 427 of the 3,505 critical statements were from that area. Several international environmental groups – the Cousteau Society, the Council of the European Ichthyological Union, the IUCN – passed resolutions condemning the airport development, but despite signs of discomfort by the central government,[27] the prefecture ignored all comments from outside the related area.

**Table 12.3** Comparison of EIAs for the NIA

| | Pacific Consultants | New Japan Marine Climate Corporation | Okinawa Prefecture No. 1 | Okinawa Prefecture No. 2 |
|---|---|---|---|---|
| Date completed | March 1981 | 1983 | September 1986 | April 1988 |
| Length | 387 pages | 562 pages | 690 pages | 800 pages |
| Site area | 130ha | 130ha | 130ha | 110ha |
| Runway length | 2,500m | 2,500m | 2,500m | 2,000m |
| EIA guidelines used | na. | na. | EIA Cabinet Decision MoC guidelines on reclamation projects | MoT technical guidelines on reclamation projects |
| Scope | Present environment Reclamation effects Operational effects | Present environment Reclamation effects Operational effects | Present environment Reclamation effects | Present environment Reclamation effects |
| Public inspection | Not allowed | Not allowed | 25 July 1986/25 August 1986 | 27 April 1988/13 June 1988 |

| Criticisms, comments | | | |
|---|---|---|---|
| — Objective description of present conditions<br>— Clear methodology<br>— Mentions difficulty of evaluation without cooperation of locals<br>— Admits potential negative effects<br>— Proposes measures for pollution control | — Methodology not clear<br>— Only conclusions, little factual data to support them<br>— No references<br>— Consideration of accuracy difficult | — All items interpreted in the developer's favour<br>— Important elements dropped<br>— No basic survey map<br>— Methodology unclear<br>— Several data inaccuracies<br>— Some assumptions are incorrect<br>— Claims for the effectiveness of pollution counter-measures unsubstantiated | — No assessment of airport operation<br>— Consideration of alternatives inadequate<br>— State of the coral in the Ryukyu Archipelago not discussed<br>— Consideration of need not sufficient<br>— Lack of references<br>— Questionable data sources<br>— Insufficient, poor quality data<br>— Ecological analysis lacking<br>— Value judgements questionable<br>— Effectiveness of counter-measures unsubstantiated<br>— Lack of safety considerations |

Source: Committee to Consider the New Ishigaki Airport Problem (October 1986), pp. 47–61; Suzuki (August 1988), pp. 1–20.

## Effectiveness of the EIA procedures

This section discusses the 1988 EIA. Figure 12.3 shows the EIA procedures used. This discussion is particularly detailed because the NIA's 1988 EIS is a prime example of how an EIA should *not* be performed. The final section summarizes how it could be improved.

**Figure 12.3**   EIA procedures for the New Ishigaki Airport

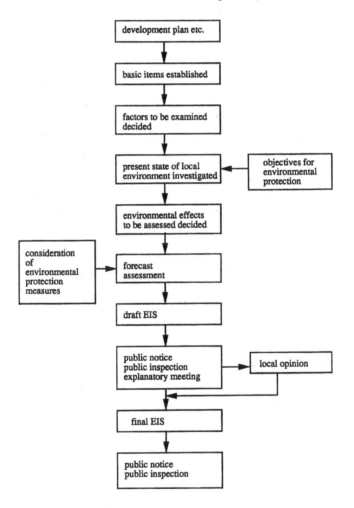

Source: Okinawa Prefectural Government (April 1988), p. 5.

*Comprehensiveness*

The EIS did not discuss whether the project is necessary, even though it has been mainly justified on economic grounds. The economic arguments for the NIA have been criticized because they do not mention the project's potential negative economic impacts on the islanders' lives:

— Opportunities for other developments such as marine or research parks may be lost.
— Provision of a jet service may not be able to close the gap between freight charges on produce coming from the Yaeyama islands and that from other competing areas. With the possible exception of tuna and rock lobster, marine products are also not expected to sell well on the mainland. The Ishigaki market will be opened more directly to mainland producers and imports may greatly increase. Greater use of air freight may also harm the shipping industry.
— Although the local construction industry may receive some spin-offs from the airport development, the main contracts for the ¥38 billion airport are expected to go to off-island companies. The local economy could become increasingly dependent on other major construction projects to prevent a boom–bust situation during the run-down of the airport construction.
— Most local hotels and tourist facilities are not expected to benefit from the airport development. Instead the big airline and tour companies, which already own a large amount of land on the island, are expected to develop their own almost self-contained resort centres.[28]

The NIA plan could thus gradually destroy the economy of the Yaeyamas as well as its natural environment. The exclusion of these impacts raises doubts about the value of a study where the contents and the importance attached to each item have been decided solely by the developer.

The EIS was also flawed by its lack of consideration of alternatives. Only seven pages of the 771-page draft EIS dealt with alternatives to the proposed site. This reflects the fact that the site was decided on before the airport's effects were assessed, and that the EIS primarily seeks to justify the plan rather than to create an environmentally acceptable one.

Table 12.4 shows the range of environmental factors considered in the EIS. Only the effects of the construction and existence of reclaimed land are considered, not those related to the airport's operation.[29] Thus the EIS did not discuss such environmentally damaging effects as the run-off of oil and other pollutants from the airport, noise and air pollution from airplanes, sewage, and the visual impact of related buildings and infrastructure. In addition, activities such as the siting of landing lights were described only briefly because they are not part of the reclamation scheme.

**Table 12.4** Environmental factors considered in the EIA for the NIA

| | Air quality | Water quality | Noise | Vibration | Marine climate | Topography | Floral fauna | Marine life |
|---|---|---|---|---|---|---|---|---|
| Existence of the reclaimed land | | x | | | x | x | | x | x |
| Construction | x | x | x | x | | | x | x |

Note: Local social conditions including population, land uses, industry, water supply and traffic were described. In addition, provisions specified in the related environmental laws were also given lip service. These include the Basic Law for Environmental Pollution Control, the Natural Environment Conservation Law, Natural Parks Law, Cultural Assets Protection Law, Wild Life Protection Designated Areas as well as local government ordinances.
Source: Okinawa Prefecture (April 1988), p.6.

The EIS also did not consider the NIA's impact in the context of the general destruction of the archipelago's reefs. Shiraho's potential as a source for the colonization and recolonization of other areas was not addressed, nor was its value as a spawning ground for commercial fishing. Finally, the EIS did not address safety considerations such as the consequences of typhoons and tsunami, even though Shiraho was once victim to a tsunami which wiped out nearly the whole village.

*Method of assessment*

The data collection techniques, assessment methods and presentation of the prefecture's EIA have been criticized as being unsystematic and incomplete, as illustrated by the following examples:

*Species diversity* Table 12.5 compares data on coral and fish species at the reef from the prefecture's EIS and from the IUCN. The EIS's lower

**Table 12.5** Species diversity at the Shiraho reef: EIS and IUCN

| | EIS | IUCN |
|---|---|---|
| Corals: | | |
| Families | 15 | 17 |
| Genera | 33 | 43 |
| Species | 79 | 100 |
| Fish: | | |
| Families | 52 | 83 |
| Species | 280 | 317 |

Source: Suzuki (Aug. 1988), pp. 13–14.

counts reflect problems with the prefecture's survey procedures. For instance the EIS coral data are based on surveys of Jan.–Mar. 1984, Nov. 1984–May 1985, Oct. 1985–Feb. 1986, and Jan.–Feb. 1987. These were all winter surveys and thus do not reflect seasonal changes. In each case the methodology and the research teams were different. Although the IUCN's surveys of Nov.–Dec. 1987 were not in-depth, their results should have been sufficient to prompt the prefecture to recheck its surveys.

Other surveys were similarly limited. The prefecture's bird surveys were carried out 9am–5pm (24 July 1987), 11am–3pm (7 Dec.) and 3pm–5pm (8 Dec.). However, migrating shorebirds can be seen in spring and autumn, and other birds are in evidence at dawn and dusk. The surveys seem to coincide more with prefectural office hours than with the task of realistically cataloguing the birds present. Local reptiles which have no special status in conservation regulations were not listed in the assessment, nor was the o-kimori, a large bat found at the site. The discovery of a rare hermit crab at Shiraho was described as accidental.

*Ecosystem requirements* The EIS did not analyse organism needs, so predicted impacts cannot be compared to impact thresholds. For instance, the EIS concluded that the NIA would destroy the coral within the reclamation area but only minimally affect that outside the area. The EIS also classified the coral into two groups, blue coral and other, and claimed that the blue coral can be conserved independently of the rest of the reef (the EA had recommended that the project be approved if the blue coral can be conserved). However, ecologists argue that the existence of blue coral hinges not only on its location but on the entire ecosystem,[30] and that the NIA's construction will destroy the marine ecosystem and kill all coral within 2–3km.

*Tidal flow and currents* Tidal flows and currents influence the impact of reclamation on the ecosystem. The EIS's current charts were based on only one survey, at two points outside the reef and one inside. No study was made of the effect of wind direction and velocity on current patterns, of the radiation stress of wave action on the shallow shore,[31] or of the effect of storm tides and typhoons on the landfill area.[32] The EIS also disregarded such factors as salinity, temperature and residence time of the water.[33] As a result the EIS is too simplistic and underestimates the changes that will occur.

*Suspended solids* The EIS estimated that the NIA's reclamation works would generate 1.27 tons/day of suspended solids. This represents a 47-fold increase over present levels. Even if only 0.1 per cent of the 3.8 million$^3$ of reclamation fill ended up on the reef, it would be enough to cover 1,900km$^2$ of coral in 2mm of sediment.[34]

*Environmental conservation measures*    The EIS claims that collecting ponds, silt nets and chemical precipitants will ensure the maintenance of EQSs. The ponds, however, have been criticized as being too small and likely to overflow during the typhoons which pass annually through Ishigaki. According to designer standards, silt nets are only 80 per cent effective. Chemical precipitants are not effective in practice,[35] and the EIS did not address their possible impact on the coral.[36] Many of the details of mitigation measures and environmental monitoring were delegated to the not-yet-established Environmental Surveillance Committee without further clarification, under the assumption that such measures will exist and work.

*Data sources*    Studies of the reef which were not commissioned by the government, such as the 1986 World Wide Fund for Nature survey or the 1988 IUCN report, were ignored. Despite the EIS's lack of references, it is obvious that many of its conclusions are based on two reports of 1985 and 1987 commissioned by the local authorities from Shoei Shirai. Mr Shirai is not a recognized coral reef ecology specialist.

### Review

Public consultation was severely restricted. There was no public participation in the 1981 and 1983 EIAs, and the site selection procedure was not open to public consultation. The 1986 and 1988 draft EISs were available for comment but under limited conditions.

The large number of opinion statements on the 1988 draft EIS showed the strength of public opposition to the project. In July 1989 fifty citizens filed a suit against Okinawa's governor at the Naha District Court demanding the refund of ¥15.2 million paid to the New Japan Marine Climate Corp. for surveys which formed the basis of the NIA's EIS. The plaintiffs claimed that the surveys were insufficient and unscientific and contained serious mistakes including the alteration of data. The outcome of this case is unknown at the time of writing but will be of great interest for the future of Japan's EIA system.[37]

Political momentum, both national and international, has formed behind the opposition groups. Some 365 members of the Osaka Lawyers Association issued an opinion statement against the proposed airport,[38] and the IUCN, World Wide Fund for Nature, and Cousteau Society have all condemned the project. However, despite litigation, participation in elections and international pressure, the NIA still looks likely to be built.

The administrative review procedures were one of the most problematic aspects of the assessment. Okinawa prefecture is both the project proponent and its EIS reviewer, and thus cannot be expected to review without bias. The governor did appoint a fourteen-member Informal

Advisory Committee of academics and engineers to comment on the 1986 EIA. The committee visited Ishigaki only once, and did not consult with local people. Their report of July 1986 merely suggested that efforts be made to prevent the fouling of water and disturbance of the sea bed during construction, and to replant the coral afterwards.

At the national level, the EIS must be reviewed by the MoC and the EA. The EA did adopt a strong nature conservation stance regarding the NIA, but as mentioned earlier has now lost its role in the decision making.

### Conclusions

The prefecture's slapdash handling of scientific information mirrors Japan's general approach to environmental issues. The so-called scientific assessment system is riddled with undefined value judgements which stem from the inherently political nature of decision making. Science is used to justify decisions, not to help make them.

In Japan there is no threshold, no agreement on when an action is too environmentally damaging to be allowed. The EQS system is useless when standards can be maintained while coral reefs are being destroyed. Under these conditions EIA becomes a mere ritual with a predetermined outcome rather than a source of information for decision makers. Conservationists can at best only delay an environmentally damaging development in the hope that a change in political or economic conditions will cause the plan to be abandoned, or that the developer produces an acceptable EIA.

Despite strong local and international pressure, no comprehensive, unbiased assessment of the NIA has taken place. The need for the airport, its siting and the consideration of alternatives were removed as quickly as possible from the forum of public debate. The EIA's terms of reference and the importance of the items to be assessed were determined by the developer. The EIA set out to prove that the project's effects will be small. All its assumptions were aimed at supporting this conclusion and any evidence to the contrary was ignored. The local authorities merely sought to justify, through the EIA system, a decision made in 1979 without public consultation. When it became obvious that the original plan was unacceptable, the prefecture simply switched sites, incurring additional costs to taxpayers and project opponents. If the EIA process had been carried out correctly, this – and future – waste of resources would have been avoided.

The case of the NIA shows that EIA is ineffective in preventing environmental destruction unless it is supported by an appropriate political climate. A development action is virtually unstoppable if the administrative agencies involved are committed to it. Especially when the developer is also the EIA reviewer, one can only expect that the outcome will be pro-development.

Rather than seeing the NIA in purely black terms, however, we will summarize possible improvements which could also apply to other EIAs. Any new EIA for the NIA should set out to do the following:

— Discuss the need for the project and alternatives to the project.
— Discuss all aspects of construction, existence and operation of the project.
— Discuss the project within the context of overall developmental and environmental trends in the area.
— Discuss the interrelations of the ecosystem and the needs of its organisms. Link impact to organism needs.
— Adapt survey methods to the ecosystem and organisms. Consider locational and temporal fluctuations.
— Include peak conditions, one-off cases and worst-case scenarios, and their probability of occurrence.
— Question the effectiveness of EQSs in preventing environmental harm.
— Compare the project's effects with those of similar previous developments.
— Buttress any assumptions made. Discuss the robustness of assumptions: if they are incorrect, how severe are the consequences?
— Keep methodology constant over surveys.
— Assess effectiveness and availability of pollution mitigation measures in practice. Do not assume future developments in technology.
— Use data from other reports for comprehensiveness, but question assumptions and limitations of that data.
— Encourage local participation in determining the importance of items to be assessed, and in ensuring that surveys are comprehensive.
— Facilitate public participation: allow EISs to be copied, hold public hearings in sufficiently large venues, make allowances for non-local comments on the EIS.
— Compensate for bias in the reviewers and investigators.

### Notes

1. Okinawa prefecture claims that the new site is 4km north of the original site, whereas environmentalists claim that it is merely 1.5km away.
2. (a) *Days Japan* (2 July 1988, in Japanese), 'Letter to Ishihara, Minister for Transport: you can save the Shiraho Sea', 9, pp. 114–22.
   (b) Suzuki, M. (June 1988), 'Shiraho news'.
   (c) Suzuki, M. (Aug. 1988), 'Shiraho news'.
   (d) Committee to Consider the New Ishigaki Airport Problem (Oct. 1986).
   (e) Citizens Group to Consider the Airport Problem (1987, in Japanese).
3. Okinawa Prefectural Government (Apr. 1988), p. 1.

4. *Ryukyu Shimpo* (15 Jan. 1985, in Japanese), Letter from Nishime Junji, Governor of Okinawa Prefecture, in reply to the Cousteau Society.
5. Committee to Consider the New Ishigaki Airport Problem (Oct. 1986).
6. Muzik (1985).
7. International Union for the Conservation of Nature (Mar. 1988).
8. On 4 Aug. 1988 an EA official at Ishigaki admitted that the quality of the coral in the Iriomote National Park Marine area (see Figure 12.1) is now below that of the Shiraho coral reef.
9. IUCN (Mar. 1988), pp. 2–3; Shiraho Protection Group (Apr. 1988), p. 13.
10. Survey carried out by the Committee of 100 for Peace and Research by Dr Shigekazu Mezaki of Mie University: Suzuki (Aug. 1988); Suzuki (Dec. 1987) 'Shiraho news'.
11. IUCN (Mar. 1988), pp. 2–4.
12. Construction of the New Amami Airport began in 1983 and it opened in July 1988. The 100ha airport has a 2,000m runway and was built on top of a coral reef. A study in April 1988 by Okinawa University showed that in an area 2–3km around the airport all of the coral is either dead or dying: *Mainichi Shimbun* (30 June 1988).
13. Committee to Study the Socio-Economic Effects of the New Ishigaki Airport (Feb. 1986), pp. 7–13.
14. Ibid., pp. 10–11.
15. *Days Japan* (2 July 1988, in Japanese), 'Letter to Ishihara, Minister for Transport: you can save the Shiraho Sea', 9, p. 121.
16. Ibid.
17. Committee to Consider the New Ishigaki Airport Problem (Oct. 1986), pp. 14–26; Osaka Lawyers' Shiraho Study Group (Mar. 1989), pp. 34–54.
18. After carrying out confidential surveys at Shiraho and the New Amami Airport, the EA embarked on a survey of the coral reefs at Ishigaki in November and December 1988. Twenty-one areas were surveyed at a cost of ¥8.1 million. The results showed that the Shiraho reef had by far the highest percentage of living coral, with 72 per cent, 52 per cent and 46 per cent at the three points surveyed. The East Karadake site had 14 per cent live coral, better than nine other areas surveyed.
19. *Days Japan* (2 July 1988, in Japanese), 'Letter to Ishihara, Minister for Transport: you can save the Shiraho Sea', 9, p. 121.
20. Suzuki (June 1988), 'Shiraho news', pp. 15–18.
21. Ibid.
22. Okinawa prefecture sent its application via the Okinawa Development Agency.
23. Suzuki (Nov. 1988), 'Shiraho news'.
24. Environment Agency (Feb. 1985).
25. Five times more than for the 1986 EIS, almost twice that for the Honshu–Shikoku bridges, and almost 40 times that for the Kansai International Airport.
26. There was no such restriction for the 1986 EIS.
27. Several politicians spoke out on 11 Feb. 1988 in support of the IUCN resolution.
28. In 1972 financial interests from mainland Japan bought about 27 per cent of Ishigaki island. Since then local people have been trying to buy back some of this land but the airport plan has resulted in renewed interest by mainland buyers.
29. International Union for the Conservation of Nature (July 1988), p. 2.
30. Suzuki (Aug. 1988), pp. 13–14.

31. This gives rise to currents very different from tidal ones.
32. The 1986 EIS noted that waves up to 15m in height break on the reef crest.
33. IUCN (July 1988), pp. 10–11.
34. Ibid., p. 11.
35. The amount of chemical coagulant used must be precisely determined in accordance with the amount of suspended solids. Coagulation cannot effectively take place if the amount of coagulant is too great or small. As the amount of suspended solids will vary considerably, it is unlikely that a suitable coagulant density can be maintained in the field.
36. Okinawa Prefectural Government (Apr. 1988). The head of the Okinawa prefecture's Agriculture, Forestry and Fisheries Dept. admitted at a meeting of the prefectural assembly that complete containment is problematic, given the present state of technology.
37. *Ryukyu Shimpo* (20 July 1989, in Japanese), 'Pay back the survey fees: lawsuit against the Governor'.
38. Suzuki (March and May 1989) 'Shiraho news'. The lawyers' report of May 1989 is entitled 'Coral for future generations: a statement of opinion in favour of reassessing the New Ishigaki Airport plan'.

# Chapter thirteen

# Trans-Tokyo Bay Highway

## Introduction

The six-lane Trans-Tokyo Bay Highway (TBH), when completed, will be 15.1km long (10.1km bridge and 5km tunnel), and will connect the cities of Kawasaki and Kisarazu (see Figure 13.1). Construction began in 1988 and will take ten years at an estimated cost of ¥1.15 trillion.[1] Under article 3 of the TBH Special Law, the Japan Highway Public Corporation (JHPC) is responsible for planning the highway and for preparing its EIA.

The fact that the highway is located in Tokyo Bay meant that it was difficult for the JHPC to define the local impact area and thus determine which EIA guidelines were applicable. At one time it was possible that the project might be subject to assessment not only under the 1985 MoC procedures for EIA, but also under those of one metropolitan government, two prefectures and two cities. This case study shows how national and local EIA procedures were coordinated to resolve this problem.

## Economic arguments for the TBH

The TBH will not be a big traffic generator: an optimistic estimate predicts that 64,000 cars/day will use it. It will reduce journey times between Kawasaki, Tokyo, Yokohama and Kisarazu by an hour at most.

The bridge's major advantage is that it will add further impetus to the development of Kanto's regional economy. It is expected to encourage the development of the Boso peninsula (Chiba prefecture) which holds much of the developable land near Tokyo. This area could provide Tokyo with land for new residences and industrial and research facilities. In its EIS, the JHPC calculated that the TBH could increase south Kanto's economy by around ¥5 trillion annually and raise the local tax base by ¥200 billion by the year 2000. However, like a two-edged blade, the TBH could either relieve Tokyo of some of the burden of its overcrowding or encourage the further concentration of activities in the area. It might also

**Figure 13.1** Trans-Tokyo Bay Highway

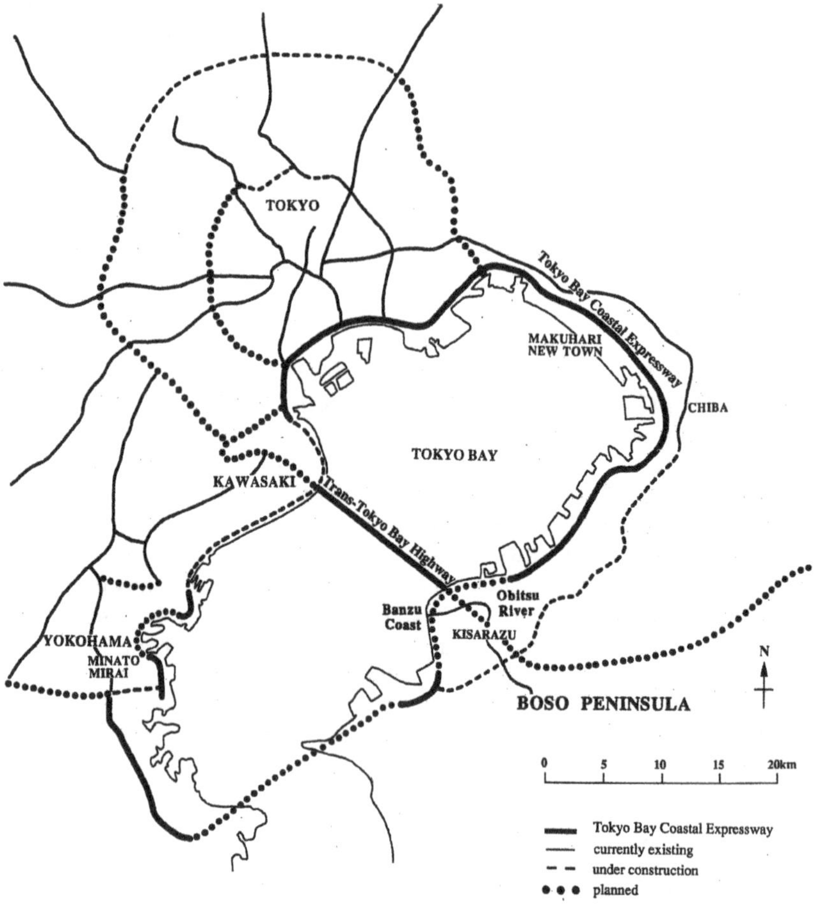

cause activities to relocate from other areas to south Kanto, resulting in a limited net gain to the regional economy.[2]

The project was strongly supported by Chiba prefecture, local politicians, economic circles and in particular Japan Steel. The latter stands to benefit from the use of steel in the bridge's construction and, as a major landowner, from the development of the Boso peninsula. Former PM Yasuhiro Nakasone was also a keen advocate of the project. He saw it as an excellent opportunity to promote the privatization of highway construction in line with the privatization of the national railways. Already in 1965 the Economic Planning Agency had selected the TBH as a possibility for privatization. The private sector, however, had been hesitant because of the huge investment required. Industrialists needed assurance on the potential returns on their investment, and in particular on the amount of traffic that would be generated. Nakasone, when elected as PM in 1982, agreed that the government would cover the risks via the JHPC.[3]

On 7 April 1986 the Diet passed a law to bring about the construction of the TBH, and five months later the Trans-Tokyo Bay Highway Company was established.[4] The MoC, the project's principal promoter, hoped that the TBH Co. would allow the vitality of the private sector to be used. The company was given responsibility to carry out the highway's basic design, explore funding possibilities, implement and maintain the project, and operate the tolls. It was also given government guarantees on financing and tax exemptions.

This approach has three advantages. First, as part of the private sector, the company can seek alternative sources of finance. Second, the JHPC can delegate many of its responsibilities for the project and therefore exclude the TBH when choosing between various priorities for future investment; in other words, it gives the JHPC a free hand to continue with its other highway plans. Third, an incentive clause ensures that the project is completed on time and is financially viable. Although the project's total costs are estimated to be ¥1.15 trillion, as with all major projects the true cost will not be known until the project is completed.

### Local authority negotiations

Although the local authorities on the east side of Tokyo Bay were enthusiastic about the TBH, those on the west side (Kanagawa prefecture and Kawasaki, Tokyo and Yokohama cities) were less so. These politically progressive administrations[5] were concerned about the potential nuisance caused by the development. Kanagawa's governor, Ichiji Nagasu, believed that many projects had a higher priority than the highway. Kawasaki city believed that the project would primarily benefit Chiba, and that they themselves did not need Chiba or the highway in order to further develop their city. Saburo Itoh, Kawasaki's mayor, was also originally against the plan. However, the local authorities gradually accepted the views of local economic circles and the national administration on the need for the highway and the need to use the space on the peninsula to their advantage,[6] especially after the MoC exerted influence on them through its city planning, highway and sewerage powers. In the negotiation process Kawasaki city was able to get the MoC to agree to provide it with direct access to the Tomei Expressway. Previously most traffic passed though Kawasaki on its horizontal axis to enter Tokyo. Kawasaki had objected to the original plan for a ring road and bridge and had asked for a new road to be built through its vertical axis, on the grounds that this would improve communications, the environment and access to its port.

Local opposition groups remained unconvinced of the need for the TBH. At one point twenty groups opposed the project. In November 1986, one month after the TBH Co. was set up, these groups merged to form the Trans-Tokyo Bay Highway Development Opposition Group. Although this group was reasonably successful in raising local awareness of the project's potential environmental damage and in pressuring the local authorities to ensure that adequate measures were taken to protect the environment, it failed in its main objective – to bring about the cancellation of the project. The group described the highway's EIA as unscientific, with pre-determined results. In particular, it claimed that the economic arguments for the TBH project were flawed. The highway's positive economic effects as predicted in the EIS relate directly to the follow-up projects planned by the MoC (i.e. artificial island developments). The fact that these projects' environmental impacts were not included in the EIS whereas their economic impacts were negates the meaning of the assessment.[7]

### EIA of the TBH

Determining the relevant EIA procedures for the TBH required extensive negotiations. With six lanes and a length of over 10km, the TBH was subject to assessment under several national and local EIA systems. However, the relationship between national and local guidelines, and between prefectural and city procedures was unclear. It was thus necessary to determine which procedures were applicable. Figure 13.2 shows how to determine which guidelines apply: the TBH was a case 3 project.[8]

The next step was to determine the relationship between national and local EIA systems. The 1984 Cabinet decision requested that local authorities 'pay due attention to conformity' with the national guidelines, but did not mention how this conformity was to be achieved.[9] Because the development was to be located in Tokyo Bay, the Tokyo Metropolitan Government, Kanagawa and Chiba prefectures, Kawasaki and Kisarazu cities, and perhaps even Yokohama could all claim to be affected by the proposed development. However, after negotiation between the parties involved it was agreed to delineate the affected area on the basis of two criteria:

(a) whether the highway was a relevant project under local authority EIA procedures; and
(b) whether the area defined under (a) corresponded to the MoC EIA guideline's definition of 'related area'.

**Figure 13.2** EIA guidelines for highway projects

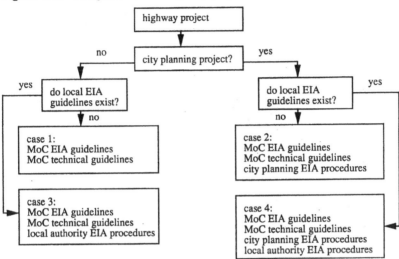

Source: Mori (August 1989), p. 196.

This eliminated Tokyo and Yokohama. Of the remaining authorities, Kanagawa prefecture's EIA ordinance was unnecessary if Kawasaki city's ordinance was used, and Kisarazu city didn't have EIA guidelines so Chiba prefecture's guidelines applied. Thus only Kawasaki's, Chiba's and the MoC's procedures were followed. Not all of the local authorities were happy with this arrangement and in no way did it solve all the problems of conformity. For example, at least six major differences exist between Kawasaki's ordinance and the MoC guidelines. In Kawasaki,

— the mayor, not the developer is responsible for announcing the draft EIS;
— the developer has to inform local people and report on the results;
— anybody, not just residents of the related area, can comment on the EIS;
— the developer must report on the local opinions received and the measures needed to reflect these opinions in the EIS;
— the mayor produces a report commenting on the EIS, opens it for public comment, and can call for a public hearing; and
— the developer must report on post-construction monitoring studies.[10]

It was agreed that the JHPC would follow the procedures specified in the MoC guidelines while Kawasaki would follow its own procedures. Where these procedures were the same only one party would carry them out, but where they differed both parties would carry out their own part.

**Figure 13.3**   Simplified EIA procedures for the Trans-Tokyo Bay Highway

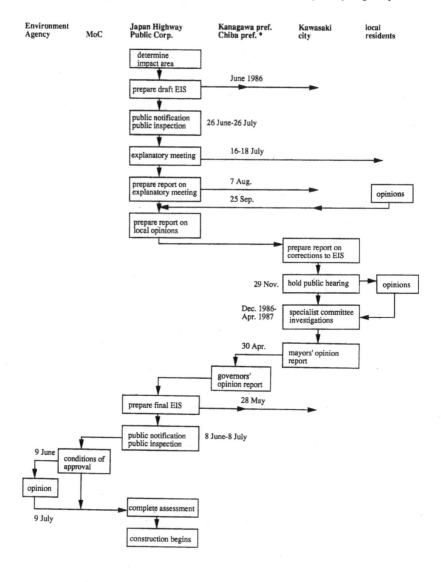

Note: * Kanagawa and Chiba prefectures' EIA procedures vary greatly, but are combined in this figure for simplicity's sake.
Source: Mori (August 1989), p. 199.

The same relationship applied to Chiba prefecture. Figure 13.3 summarizes this procedural flow. The JHPC was primarily responsible for notifying the various parties involved. These functions vary little between the local and national guidelines and were thus easier for the developer to carry out. The local authorities, on the other hand, were responsible for collecting and presenting opinions, the procedures for which vary greatly between the national and local systems.

**Effectiveness of the EIA**

Kawasaki city and Kanagawa prefecture have some of the most comprehensive EIA systems in Japan. They include requirements for setting up specialist committees, preparing reports on alterations to the EIS, holding public hearings and fairly widespread public participation. Their only real weakness is the lack of an alternatives requirement which means that there was no real comparative basis on which to analyse the impact of the TBH.

*Comprehensiveness*

The range of environmental impacts considered in the assessment is shown in Table 13.1. This range is comprehensive and the JHPC did its best to carry out a thorough assessment of these items. However, two problems remain. First, the items to be examined were derived from the MoC guidelines on EIA. No attempt was made at scoping or tiering to identify which impacts were likely to be most significant or of greatest interest to local groups. Second, safety and ecosystems were not discussed. There was no assessment of the potential dangers of accidents involving shipping during the highway's construction and operation. Tokyo Bay is heavily congested and at least the impact of additional construction traffic should have been assessed. Furthermore, although the EIS deals with land flora and fauna and marine life, no attempt was made to view the ecosystem as a whole. This omission angered local opposition groups which claimed that the highway and related developments endanger the last remaining tidal flats and salt marshes in Tokyo Bay.[11]

*Public participation*

Public participation in the TBH assessment was widespread. The main form of participation was through the completion of opinion statements. In total, 1,770 statements were received, of which 1,746 were opposed to the project. In addition, 154 people attended Kawasaki city's public hearing and thirteen were allowed to speak. The main points raised at the hearing were that $NO_2$ levels in the bay area were already high and would

**Table 13.1** Environmental impacts considered in the EIA for the TBH

| | Air pollution | Water pollution | Noise | Vibration | Ground configuration | Land plants | Land animals | Marine life | Scenery |
|---|---|---|---|---|---|---|---|---|---|
| *Construction* | | | | | | | | | |
| Man-made island | x | x | | | | | | x | |
| Tunnel | | | | | | | | | |
| Bridge | x | x | x | x | | | | x | |
| Roads | x | | x | x | | x | x | | |
| *Operation* | | | | | | | | | |
| Man-made island | | x | | | | | | x | x |
| Tunnel | | | | | x | | | | x* |
| Bridge | | x | | | | x | x | x | x |
| Roads | | | | | | x | x | | |
| Traffic | x | x | x | x | | | | x | |

Note: * = air ventilation towers.
Source: Japan Highway Public Corporation (1987), pp. 41.

worsen if the project went ahead, that the tidal flats would be lost, and that water quality would continue to decline. The widespread criticism of the project did not, however, delay the assessment significantly and at no time did the assessment's outcome seem in doubt.

### Administrative review

The draft EIS for the TBH was completed in June 1986. The review process from that time on took one year. In total, the EIS was reviewed by the governors, mayors and various committees of eleven local authorities. Investigations were most extensive in Kawasaki, with eleven committee meetings taking place between November 1986 and April 1987, and a one-day public hearing on 29 November. Chiba prefecture held eight committee meetings between August and December 1986, and a public hearing on 11 November.[12]

Chiba was concerned about the water quality in Tokyo Bay, conservation of the Obitsu River and noise/vibration from the bridge, but concluded that the EIS was *oomune datou* (generally OK). Kawasaki city was concerned with traffic volume on the connecting roads, with the NOx and suspended particulate matter generated, and with the water quality of Tokyo Bay.[13]

Approximately six months of the review process were taken up by Kawasaki city's in-depth analysis of the EIS. Specialist committees reported to the mayor on these issues in April 1987. The review process resulted in the JHPC implementing several expensive measures to contain emissions.[14] Had that analysis not taken place, the review process under the national guidelines could have taken a mere six months and would have been far less rigorous.

The final EIS was 962 pages long, with another 312 pages of supporting data, and was submitted to the MoC by the JHPC on 9 June 1987. Although the local authorities attempted an unbiased review of the EIS, the same cannot be said of the MoC. The MoC has the final say with regard to the assessment and licence approval, but since the JHPC is one of its organs and the MoC itself was one of the project's major promoters, one cannot expect it to review the assessment without bias. The MoC approved the development one month after it received the EIS.

Finally, the EA commented on the assessment. It raised seven points related to the ventilation towers, traffic, water quality, conservation of the tidal flats, landfill materials and monitoring measures.[15]

### Conclusions

The EIA for the TBH was generally good, but more by accident than by design. It was reasonably comprehensive, public participation was encouraged and review was thorough. However, these factors were due to Kawasaki's, not the MoC's, EIA procedures. Had the TBH not fallen within the area of influence of a local authority with strong EIA procedures, its assessment is likely to have been much less thorough and ameliorative measures may well not have been implemented. The validity of the TBH's EIA is also debatable because of its inaccurate balancing of economic benefits and environmental costs.

This case study shows that the 1984 Cabinet decision on EIA did little to bring about uniformity in Japan's EIA systems. Today's system is more complicated, more duplicative and more wasteful of local authority time than that of pre-1984. Major development projects like the TBH which are covered by more than one EIA system require extensive negotiations between the various organizations involved in order to ensure that the assessment is acceptable to all parties. The end result can be a procedural hodge-podge rather than an organized and easy-to-understand procedural flow.

### Notes

1. Trans-Tokyo Bay Highway Co. (May 1987).
2. Japan Highway Public Corporation (no date).
3. Interview with Professor Fumio Takeda of the Highway Research Group, 9 Aug. 1989.
4. Oshima (1989), pp. 994–5.
5. Socialists and Communists are represented in their assemblies.
6. Interview with Professor Takeda, 9 Aug. 1989.
7. *Environment Assessment Handbook* (1988), pp. 187, 204–6, 269–71.
8. Mori (Aug. 1989), pp. 195–201.
9. On 12 Dec. 1984, the EA informed governors and mayors of designated cities that the project developers were required to ensure that conformity with procedures was achieved.
10. Mori (Aug. 1989), p. 198.
11. The highway will connect to the Banzu coast near an area which contains the bay's most extensive tidal flats and the only remaining salt marsh. The local fishing cooperative received ¥24.7 billion in compensation for the loss of fishing rights around the Banzu area. Conservation of the marsh was not ensured: Short (1989), pp. 1, 6–7, 16.
12. This compares with the 340-day public inquiry held for the Sizewell 'B' nuclear power station in the UK, which generated 16 million transcripted words and cost the Central Electricity Generating Board at least £15 million (¥3.4 billion) and the opposition £750,000 (¥0.2 billion); Elkinton and Burke (1989), p. 151.
13. *Environment Assessment Handbook* (1988), pp. 187, 204.
14. The cost of these measures, details on their nature, and their likely effectiveness was not addressed in the EIS.
15. Environment Agency (July 1987).

# Chapter fourteen

# Kyoto Second Outer Circular Route
*by Toshio Hase*

### Introduction

This case study concerns the MoC's construction of a national highway through the southwestern section of Kyoto city (see Figure 14.1). Kyoto prefecture's governor decided that the project should proceed, with strong guidance by the MoC. Although EIA is now a part of the construction decision-making process, the government uses it to enhance the project's image in an attempt to minimize public participation. Environmental protection and the opinions of the residents who will live near the highway seem to be minor elements in the construction decision.

Japan's EIA system functions in such a way that development is always more respected and the environment is given only lip service. Kyoto's second outer circular route appears to be no exception.

### Decision on the highway project

Oharano is situated at the extreme west of the Kyoto basin. Hills extend like green walls on one side, while the other side is edged by a new town and a bamboo forest. Most of the land is agricultural or woodland. Oharano was annexed administratively by Kyoto city in 1959. The area is designated by the city as an urbanization control area (see Chapter 4, 'Land use plans'), where land use changes are prohibited in principle.

On 2 September 1988 Kyoto's local newspaper printed an announcement by the MoC that stated that a ¥100 billion ($714 million) highway would be built through Oharano.[1] The highway would be a national project, part of the Fourth Comprehensive National Development Plan (see Chapter 4, 'National development plans'). It would be a four-lane motorway, 15.7km in length, which would connect two important national routes and pass through six municipalities in Kyoto prefecture. The announcement said that an explanatory meeting would be held in each municipality towards the end of September. The MoC planned to start construction in April 1989.[2] For the local people this sudden one-sided decision was like thunder from the blue sky.[3]

**Figure 14.1** Kyoto Second Outer Circular Highway

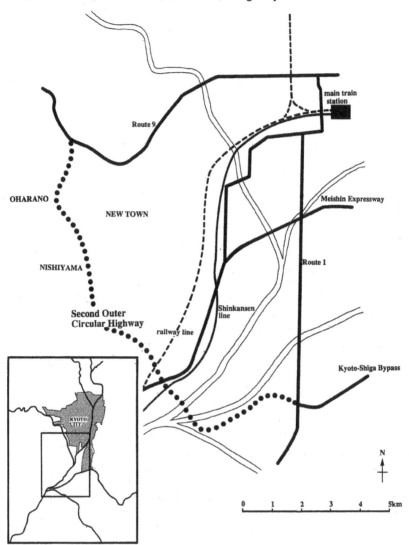

## Reactions of local people

In Oharano there exists an autonomous communal organization ('the Community') which is composed of almost all the households. The Community has a representative council, a bureau and a president. All households pay a membership fee. It is a voluntary but semi-official body which has close relations with the municipal government.

The Community considered the highway proposal to be an urgent matter. It established a special committee on the highway problem in September 1988, with ninety-two members. The chairman was the Community's president, and other members were leaders of various associations or of the locality.

The committee met at night to discuss means of opposing the project. The Community's president went to the City Hall and the prefectural office several times to express the concerns of the local people and to demand explanations. Petitions were signed and presented to the mayor and governor. However, the Community leaders did not bring in outside help or ask for specialists' advice. The Community maintains good relations with city and prefectural authorities. It tried to negotiate with the authorities, expecting special favours such as supplementary public investments in the area in return for their acceptance of the project.

Two types of residents live in Oharano: those who moved there recently to live in newly built houses, and those who have lived there for generations. The ratio is half and half. The former are people from the city, who have more liberal ideas, while the latter were originally farmers and land owners. The committee was composed mostly of the latter type.

The former, who would be more directly affected by the proposal, felt that a semi-official organization such as the Community would not be effective enough to prevent the project from going ahead. In October 1988 100 people formed a movement which was named the Association for the Protection of Nishiyama's Nature and Beauty.[4] This new movement was less conservative and more active than the Community: it felt that democracy in Japan was not yet sufficiently established through contacts with local governments, and so began printing newsletters, participating in unions, and forming links with other similar movements. The Community distanced itself from this new movement, although the movement tried to involve the Community.

### EIA and public participation

Neither Kyoto city nor Kyoto prefecture had an EIA regulation in 1988 when the project was announced.[5] The only rule applicable was the Cabinet decision on EIA. In this case the MoC had jurisdiction over the city plans according to which the highway is constructed. Thus the proposed highway was subject to the MoC's EIA procedures.

The MoC's proposal required Kyoto prefecture's governor to change Kyoto's plans. The governor determines city plans and conducts an EIA when city plans are changed. The City Planning Law, which stipulates procedures for establishing city plans, requires that proposed route plans and an EIS must be made publicly available for two weeks, and that the residents and mayors must be consulted in the decision-making process. A public hearing may be held if the governor finds it necessary. Figure 14.2 summarizes the flow of the decision-making process.

Residents blocked the first two explanatory meetings, since they knew that once an explanatory meeting is over, the government can go on to the next step in the EIA procedures.[6] The third and last explanatory meeting concerning the highway proposal was held on the evening of 21 May 1989 in a primary school auditorium, and was attended by 150 people. Local people tried in vain to stop it and demanded other meetings. They asked for more detailed explanations, but the responses were ambiguous or pointless. The meeting ended after midnight with cries and yells.

The next step was the public showing of the project plans and the EIS for two weeks. The EIS was composed of eight chapters, totalling 150 pages. Only the proposed route was considered; no alternative route was mentioned. The report concluded that the project would not seriously affect the scenery, noise, air, water or vibrations. The plans and the EIS were made public from 30 May to 13 June 1989 in three government

**Figure 14.2** Decision-making process for the Kyoto Second Outer Circular Route

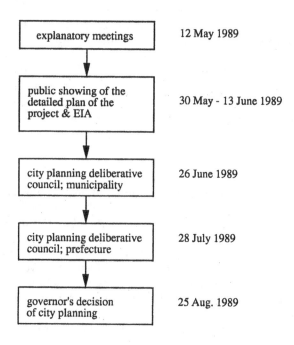

offices during office hours. No photocopy machine was available.

During these two weeks people could present opinions and comments to the governor. The opinions would be summarized and sent to the deliberative council on city planning for reference. As the Community and the movement appealed to residents and other local associations to present opinions to the prefecture, 784 were submitted. Only seven were in favour of the development while 750 were opposed; 137 mentioned the problem of air pollution, 167 the possibility of noise, and 744 were worried about the destruction of scenery; 696 suggested that the highway should pass through a tunnel in the Oharano area in the hope of avoiding noise problems. The presentation of opinions during the two-week period was only one method for expressing people's thoughts. However, the number of opinions presented shows very high interest in the project.

In addition, a member of the movement tried to meet some of the council members to convey grievances of the local residents, but he was told that the names and meetings of the deliberative council were secret.[7]

No one knows how the residents' opinions were dealt with in the council session. The governor simply approved the initial project proposal in August 1989.

The difference between the Community and the Nishiyama movement was clearly evident when the governor decided to allow the project to proceed. Upon being informed of the governor's decision, the Community's president admitted that the Community had failed in its opposition, and proposed to finish its activities in September 1989. However, most other committee members opposed the dissolution of the committee, since there would be more problems in the future when construction actually started. The Nishiyama movement was not disappointed: the governor's quick, one-sided decision was expected. The movement will increase its membership and make more appeals for public support.

### Conclusions

What is the meaning of the presentation of 784 comments and opinions by local residents? Or of the petitions filed in the City Hall and prefectural office? The time it took the governor to approve the project was 358 days. The highway project was given priority and the residents had almost no influence over the governor's decision. The MoC's EIA procedures worked perfectly in their favour in this case.

An EIA is now obligatory in Japan when the government decides to construct a highway. However, it seems that development is always given priority over the environment. The adoption of an EIA directive seems to have no significant meaning, nor does it have a role in protecting environmental quality. Instead the nominal EIS is used by pro-development ministries to justify the proposed project.

There are two reasons for this phenomenon. First the government's pro-development ministries such as the MoC and MITI have never lost their influence and power *vis-à-vis* other minor agencies such as the EA. The dominance of the LDP for the past forty years has never allowed any change in the government's pro-industrial policies. Second, EIA was adopted in Japan in such a way that important, large development projects would never be shown to have significant impacts. Unless this was assured, EIA would never have been accepted due to the government's development-oriented structure.

The government's pro-development agencies dominate the land and the sea, and Japan has no effective policy to stop the further deterioration of the once-green country.

### Notes

1. *Kyoto Shimbun* (2 Sep. 1988).
2. Ibid.
3. Old Chinese and Japanese proverb.
4. Nishiyama ('west mountain') is the name of the hills in Oharano ('great plain'). The proposed highway penetrates the edge of Nishiyama.
5. Kyoto prefecture's EIA guidelines became operative in Sep. 1989. Even without an EIA regulation, Kyoto city has had to conduct an EIA at least twice in the past due to the demands of environmental movements.
6. *Asahi Shimbun* (24 May 1989).
7. *Asahi Shimbun* (20 August 1989).

# Chapter fifteen

# Recommendations

Environmental reform in Japan in the past twenty to thirty years has emphasized environmental standards, pollution control and other *post-facto* technical measures rather than anticipative measures for the prevention of environmental degradation. Concern has been focused primarily on human health impacts and only minimally on wider environmental protection or ecosystems. Public participation has not been encouraged, and development proposals have continued to carry more weight than those for environmental protection. The reforms that have taken place left the decision-making process virtually unchanged.

The very existence of a strict pollution control system in Japan, however, may still have a significant role in that it requires that information be gathered and that some thought be given to the environmental consequences of development projects. This may cause attitudes to gradually change, which in turn may cause people to become more aware of the need to balance economic growth and environmental protection. This chapter presents suggestions for improving environmental policy and impact assessment in Japan to speed up this process.

First, Japan's decision-making process needs to be improved so that environmental matters are given adequate consideration:

— continue developing environmental policy and environmental standards;
— strengthen the authority of government bodies concerned with environmental protection (i.e. the EA and local authorities);
— increase coordination of environmental protection, planning and decision-making procedures;
— further develop environmental monitoring;
— establish accessible, central environmental data banks;[1]
— encourage further research and development of techniques for environmental assessment and protection; and
— promote widespread environmental education.

Second, as noted in Chapter 8, impact assessment should be applied at both the policy and the project level. This would help to further the integration of environmental considerations into decision making:

— implement REMPs at the local authority level, and
— require ministries to submit EIAs along with policy proposals to the Diet.

Third, EIA should be required for all projects and policies which could have a significant impact on the environment:

— Base the determination of whether an EIA is required on whether a policy/project is expected to have a significant environmental impact. A list of policies and projects subject to EIA should be drawn up by the Environment Agency in consultation with other administrative bodies and the public.
— Some of the projects which should be subject to EIA, but are not at present, are:

> refuse combustion plants
> radioactive waste disposal sites
> waste storage sites
> all airports (not just those with runways of >2,500m)
> all railways (not just Shinkansen)
> pipelines and power lines
> urban growth and urban renewal areas
> all industrial areas
> afforestation
> open cast mining
> drilling
> groundwater pumping
> production, storage, and transport of hazardous substances.

Fourth, the responsibility for EIA should be centralized rather than distributed between several government agencies:

— require all agencies to prepare EIAs for projects under their jurisdiction which could have a significant environmental impact, based on guidelines drawn up by the EA;
— give the EA authority to coordinate the assessment of projects which do not require assessment at present, and to arbitrate inter-agency disputes concerning EIA;
— introduce a comprehensive environmental licence for developers, and require that the licence application must include an EIS; and

— allow developers to appeal to the EA against the need to prepare an EIA.

Fifth, national and local EIA procedures should be coordinated, particularly with regard to public consultation:

— coordinate the subject developments, timescales and procedural flows of all EIA systems currently in existence; and
— establish uniform EIA procedures for all local authorities, keeping in mind the effectiveness of those systems already in existence. Local authorities should deal with specific projects whereas the national system should deal with general policy issues.

Sixth, the procedural requirements of EIA should be strengthened:

— make the system mandatory;
— introduce scoping;
— produce and review the EIS at the earliest time possible;
— include timely and clear announcement of EIS and public participation requirements;
— allow anyone wishing to express an opinion regarding an EIS to do so;
— convene an independent panel of experts to review EISs;
— convene an inquiry if objections to a project are numerous;
— guarantee at least one occasion for those interested to meet the developer and administrative authority;
— make public the reaction of the developer and administrative authority to comments received, and the justifications for the final decision; and
— include monitoring procedures.

Finally, the method of assessment and EIA preparation should be improved:

— encourage training of personnel responsible for EIA;
— stress analysis of impact and significance rather than description of pollution source and receiving environment;
— stress quantitative, statistically based predictions which can be tested through monitoring;
— require simulation models to be accompanied by a discussion of possible biases, assumptions made, and how the model results would change given different assumption;
— set length limits on EIAs;
— encourage EIAs to be written clearly and without jargon, and to include a non-technical summary.

One could argue that these recommendations are unrealistic, that they have already been considered and rejected, and that they could not be effectively implemented in the US or EC, much less Japan. However, the fact that all of these recommendations can still be made highlights just how far Japan's EIA system still is from being an effective tool for environmental protection.

There is ample evidence, both in Japan and elsewhere, that environmental protection procedures in general, and EIA in particular, are cost-effective.[2] It is relatively easy to impose time limits on procedures, encourage their efficient management and coordination, and introduce cost-bearing measures to cover the additional administrative effort required. Local authorities are already making a commendable effort to widen EIA to include the concept of carrying capacity and extensive monitoring.

However, no major improvements in Japan's national system of environmental policy are expected in the foreseeable future. The hegemony of interest groups which favour economic development and the powerlessness of those which favour environmental protection will continue to undermine measures to integrate meaningful environmental policies into the administrative planning process. The long-term environmental consequences of such a situation are far-reaching both within and outside Japan, as its international influence grows and its economic model is copied elsewhere.

It has often been argued that only an economically healthy nation can afford to protect its natural environment. However, environmental destruction on a large scale will inevitably result in the curtailment of wealth creation. The short-term pursuit of economic wealth above all else can only result in its own long-term destruction. Our targets should be, after all, sustainable.

### Notes

1. These recommendations are based on a study concerning the possibility of introducing EIA into Holland: Twijnstra Gudde (Feb. 1979).
2. Cook (1979).

# Appendices

# Environmental standards

## Air pollution

Main Law:     Air Pollution Control Law
Related Laws:  Road Transport and Motor Vehicle Law
               Road Traffic Law
               Mine Safety Law
               Electric Power Industry Law
               Gas Industry Law etc.

*Air Pollution Control Law*

*Soot and smoke*
(1) Sulphur oxides
    (a) Regulation of K-values, specified values set according to $SO_2$ volume emitted and stack height (with regional variations).
    (b) Fuel use regulation– area designation
                                – seasonal designation
                                – sulphur content in fuel
    (c) Total emission control applied to specified factories in certain areas.
(2) Soot and dust: standards differ from 0.05 to 0.07g/Nm$^3$ according to type and size of facility.
(3) Hazardous substances: emission standards differ with type and size of facility and with local authority area.

| | |
|---|---|
| Cadmium compounds | 1.0 mg/Nm$^3$ |
| Lead compounds | 10–30 mg/Nm$^3$ |
| Fluorine compounds | 1.0–20 mg/Nm$^3$ |
| Hydrogen chlorine | 80–700 mg/Nm$^3$ |
| Chlorine | 30 mg/Nm$^3$ |
| Nitrogen oxides | |
|     Gas boiler | 60–150 ppm |
|     Liquid firing boiler | 130–280 ppm |

Solid firing boiler          450–550 ppm
Metal heating furnace     100–200 ppm

*Dust*  (particulate matter generated by mechanical means): standards are enforced during the construction and operation of facilities. No established emission standards.

*Automobile exhaust gas*:  permissible levels set for carbon monoxide, hydrocarbons and nitrogen oxides according to vehicle and engine type, and year of production.

*Specified substances*:  twenty-eight chemical substances are designated such as ammonia, hydrogen, cyanide, etc. No emission standards.

*Environmental quality standards for ambient air*

| | | |
|---|---|---|
| Sulphur dioxide ($SO_2$) | Daily average of hourly value | 0.04 ppm |
| | Hourly value | 0.1  ppm |
| Carbon monoxide (CO) | Daily average of hourly value | 10   ppm |
| | Average hourly values in eight consecutive hours | 20   m |
| Suspended particulates | Daily average of hourly value | 0.10 mg/m$^3$ |
| | Hourly value | 0.20 mg/m$^3$ |
| Nitrogen dioxide | Daily average of hourly value | 0.04–0.06 ppm |
| Photochemical oxidants | Hourly value | 0.06 ppm |

Notes: Measurement practised at places which represent the ambient condition. This excludes designated industrial areas, driveways and other areas where ordinary civil life is not carried out.
   Suspended particulate matter is airborne particles of diameter <10 micrometer.
   Photochemical oxidants are oxidizing substances (e.g. ozone, peroxyacetyl nitrate) produced by photochemical reactions.

## Water pollution

Main Law:       Water Pollution Control Law
Related Laws:   Sewerage Law, River Law
                Marine Pollution Control Law
                Port Regulation Law
                Mine Safety Law
                Hazardous Substances Control Law
                Agricultural Soil Pollution Prevention Law
                Seto Inland Sea Environmental Conservation Law
                Regulations for waste treatment and disposal
                International conventions on marine pollution control
                    and ocean dumping

*Water Pollution Control Law*

(1) The law stipulates standards for effluent water discharged into public water areas.
(2) Public water areas are defined as water areas devoted to public use, including rivers, lakes, harbours, coastal seas and water lines connected to them (sewerage treatment lines are excluded as they are covered by the Sewerage Law).
(3) Effluent standards.
    (a) Human health related (in mg/l):

| | |
|---|---|
| Cadmium | 0.1 |
| Cyanide | 1.0 |
| Organic phosphorous | 1.0 |
| Lead | 0.5 |
| Hexavalent chromium | 0.5 |
| Arsenic | 0.5 |
| Total mercury | 0.005 |
| Alkyl mercury | not detectable |
| PCB | 0.003 |

    (b) Environment related (in mg/l unless otherwise specified)

| | |
|---|---|
| pH | 5.0–9.0: effluent to coast |
| | 5.8–8.6: other |
| BOD | 160 (daily average 120) |
| COD | 160 (daily average 120) |
| Susp. solids | 200 (daily average 150) |
| n-Hexane extract | 5 mineral oil |
| | 30 animal and vegetable oil |
| Phenols | 5 |
| Copper | 3 |
| Zinc | 5 |
| Dissolved iron | 10 |
| Dis. manganese | 10 |
| Chromium | 2 |
| Fluorine | 15 |
| Coliform bact. | 3,000 MPN/100ml |

(4) Special regulations for semi-closed water areas.
    (a) Total emission control for COD in designated areas: Seto Inland Sea, Tokyo Bay and Ise Bay.
    (b) Reduction of discharge of phosphorous and nitrogen: Seto Inland Sea, Lake Biwa and other lakes.

*Environmental quality standards for water: human health related (in mg/l unless otherwise specified)*

| | |
|---|---|
| Cadmium | 0.01 |
| Cyanide | not detectable |
| Organic phosphorous* | not detectable |
| Lead | 0.1 |
| Hexavalent chromium | 0.05 |
| Arsenic | 0.05 |
| Total mercury | 0.0005** |
| Alkyl mercury | not detectable |
| PCB | not detectable |

* Includes parathion, methyl and EPN.
** Based on the yearly average value.

*Environmental quality standard for water: environment related*

| | pH | BOD (max.) | COD* (max.) | Suspended solids (max.) | Dissolved oxygen (min.) | Coliform bacteria** (max. MPN/100ml) |
|---|---|---|---|---|---|---|
| **River** | | | | | | |
| AA | 6.5–8.5 | 1 | — | 25 | 7.5 | 50 |
| A | 6.5–8.5 | 2 | — | 25 | 7.5 | 1,000 |
| B | 6.5–8.5 | 3 | — | 25 | 5 | 5,000 |
| C | 6.5–8.5 | 5 | — | 50 | 5 | — |
| D | 6.0–8.5 | 8 | — | 100 | 2 | — |
| E | 6.0–8.5 | 10 | — | —a | 2 | — |
| **Lake** | | | | | | |
| AA | 6.5–8.5 | — | 1 | 1 | 7.5 | 50 |
| A | 6.5–8.5 | — | 3 | 5 | 7.5 | 1,000 |
| B | 6.5–8.5 | — | 5 | 15 | 5 | — |
| C | 6.0–8.5 | — | 8 | —a | 2 | — |
| **Sea** | | | | | | |
| A | 7.8–8.3 | — | 2 | —b | 7.5 | 1,000 |
| B | 7.8–8.3 | — | 3 | —b | 5 | — |
| C | 7.0–8.3 | — | 8 | —b | 2 | — |

* COD measured with potassium permanganate method.
** MPN/100ml: most probable number in 100ml.
a Floating matter/garbage should not be observed.
b n-Hexane extracts should not be detectable.

*Standards for nitrogen and phosphorous concentrations in lakes (in mg/l)*

| | Nitrogen | Phosphorous |
|---|---|---|
| Class   I | 0.1 | 0.005 |
| II | 0.2 | 0.01 |
| III | 0.4 | 0.03 |
| IV | 0.6 | 0.05 |
| V | 1.0 | 0.1 |

## Noise

Main Law:        Noise Regulation Law
Related Laws:  Vibration Regulation Law
                       Road Transport and Motor Vehicle Law
                       Road Traffic Law

*Noise Regulation Law*

(1) Noise levels for specified factories.
(2) Noise levels for specified construction work.
(3) Noise levels for automobiles.
(4) Noise levels for late evening/midnight business activities (area for regulation specified by local authority).

*Environmental quality standards for noise*

| Category | Daytime | Morning and evening | Night |
|---|---|---|---|
| Area AA | 45 | 40 | 35 |
| A | 50 | 45 | 40 |
| B | 60 | 55 | 50 |

AA — areas where quiet is especially needed, e.g. convalescent facilities.
 A — areas used mainly for residential purposes.
 B — areas used considerably for residential, but also for commercial and industrial purposes.
Different standard values are established for main roadside areas.

# Environment-related laws

## Chronology of enactment

| Year | Month | |
|------|-------|---|
| 1946 | | Law of Special City Planning for Postwar Reconstruction. |
| | 11 | Constitution boosts local government powers. |
| 1948 | 7 | Agricultural Chemicals Control Law. |
| | 7 | Hot Springs Law. |
| 1949 | | National Comprehensive Land Development Law. |
| | 5 | Mine Safety Law. |
| | 8 | Tokyo enacts Industrial Pollution Control Ordinance with clause for 'smooth economic development'. |
| 1950 | 5 | Law for the Protection of Cultural Property. |
| | 12 | Hazardous Substances Control Law. |
| 1951 | 6 | Forestry Law. |
| 1952 | | Electric Power Sources Development Promotion Law. |
| 1953 | | Public Cleansing Law. |
| 1954 | | Revenue from gasoline tax earmarked for highways. Osaka Pollution Prevention Ordinance. |
| | 1 | Tokyo Noise Law. |
| 1955 | 8 | MITI develops first petrochemical complex at Yokkaichi. |
| | 10 | Tokyo Smoke and Soot Prevention Law. First national five-year economic plan sets 5 per cent growth rate. |
| 1956 | | Minamata disease identified. Capital Region Development Law marks beginning of regional planning. New City, Town and Village Development Law. |
| | 4 | City Parks Law. |
| | 6 | Industrial Water Law combats ground subsidence. |

| Year | Month | |
|------|-------|---|
| 1957 | | Law of Special Measures for Taxation provides incentives to invest in pollution control equipment and relocate factories. |
| | 6 | Natural Parks Law. |
| 1958 | | MHW drafts Living Environment Pollution Prevention Standards Act, but never enacted. |
| | 4 | Sewerage Law. |
| | 12 | Public Water Zone Conservation Law. |
| 1959 | 3 | Law for Industrial Siting Restrictions in Developed Urban Areas of the Capital Region. |
| | | Standards for Permitting Conversion of Farmlands. |
| 1960 | | National Ten-Year Income Doubling Plan. |
| | | Law for Promotion of Water Resource Development. |
| | | Twenty-one regions designated as 'special industrial development regions'. |
| | | Regional development plans try to redistribute population and national income. |
| | | Basic Law for Agriculture. |
| | 12 | Road Traffic Law. |
| 1961 | | Law for Promotion of Water Resource Development. |
| 1962 | | First National Comprehensive Development Plan. |
| | | Smoke and Soot Control Law. |
| | | New Industrial Cities Development Act. |
| | 5 | Law Concerning Regulation of Groundwater Pumping attempts to prevent subsidence. |
| | 5 | Law Concerning Preservation of Trees for the Conservation of Scenic Beauty of Cities. |
| 1963 | | Mishima-Numazu's residents protest against designation as special industrial development region. |
| | | MITI sets up Industrial Pollution Division. |
| | | Coastal Fishery Promotion Law. |
| | | Kinki Region Development Act. |
| | 4 | Special Law for Completion of Public Sewer Facilities. |
| 1964 | | MITI/MHW carry out first large-scale EIA at Mishima-Numazu. |
| | | Act for Promotion of the Industrial Development of Special Areas. |
| | | MHW sets up Environment Pollution Control Division. |
| | | Minamata disease in Niigata prefecture. |
| | | Basic Law for Forestry. |
| | 7 | River Law. |
| | 7 | Law to Restrict Industrialization of Urban Areas in |

| Year | Month | |
|------|-------|---|
| (1964) | | Kinki Region. |
| | 7 | Electric Industries Law. |
| 1965 | | Diet organizes Special Committee for Industrial Pollution Control. |
| | | MITI starts industrial pollution survey. |
| | 5 | Kinki Region Consolidation Plan. |
| | 6 | Pollution Control Service Corporation Law. |
| 1966 | | Chubu Region Development Act. |
| | 1 | Law Concerning Special Measures for the Preservation of Historical Natural Features of Ancient Cities. |
| | 6 | Capital Region Green Area Protection Law. |
| | 7 | CO emission standards for automobiles. |
| | 9 | 'Itai-itai' disease. |
| 1967 | 7 | Law for the Development of Conservation Areas in the Kinki Region. |
| | 7 | Law Concerning Preservation of City Areas, City Development and Conservation Areas in the Chubu Region. |
| | 8 | Ocean Oil Pollution Control Law. |
| | 8 | Law Concerning Prevention of Noise in Areas around Public Airports. |
| | 8 | Basic Law for Environmental Pollution Control. |
| 1968 | | Kanemi rice oil poisoning. |
| | 6 | Air Pollution Control Law replaces Soot and Smoke Law; K-values set for $SO_2$. |
| | 6 | City Planning Act establishes Urbanization Promotion Areas and Urbanization Control Areas. |
| | 6 | Noise Regulation Law for new factories and construction. |
| | 11 | Automobile exhaust standards. |
| 1969 | | Second National Comprehensive Development Plan aims to develop a national rapid transportation network. |
| | | Agricultural Promotion Areas Act designates APAs, zones agricultural land. |
| | | Urban Fringe Agricultural Area Improvement Act. |
| | 2 | EQS for $SO_2$. |
| | 12 | Law concerning Relief Measures for Victims of Pollution-Related Diseases. |
| 1970 | 4 | Cabinet adopts EQS for water pollution. |
| | 6 | Pollution Disputes Settlement Law. |
| | 7 | Prime minister organizes Environmental Pollution Combat Headquarters. |

| Year | Month | |
|------|-------|---|
| (1970) | 11 | Special Law for Relief Measures for Victims of Pollution-Related Diseases revised. |
| | 12 | 'Pollution Diet': |

—Basic Law revised, harmony clause deleted.

—Pollution Control Works Cost Allocation Law.

—Road Traffic Law amended to cope with traffic pollution.

—Noise Regulation Law revised to control automobile noise.

—Waste Disposal and Public Cleansing Law.

—Sewerage Law expanded.

—Ocean Oil Pollution Control Law expanded to Marine Pollution and Maritime Disaster Prevention Law.

—Water Pollution Control Law.

—Air Pollution Control Law revised to include EQS for more substances and set auto emission standards.

—Agricultural Soil Pollution Law.

—Agricultural Chemicals Control Law amended to include assessment of environmental toxicity.

—Health-Related Pollution Crime Law.

—Natural Parks Law amended.

—Hazardous Substance Control Law amended to include transport safety.

| | 12 | Phase 1 of regional pollution control programme begins. |
| 1971 | | Power Stations Thermal Power Facilities Standards revised (MITI). |
| | | EQS and effluent standards for water pollution set. |
| | 5 | Cabinet sets EQS for noise. |
| | 5 | Law Concerning Special Governmental Financial Measures for Pollution Control Projects. |
| | 6 | Offensive Odour Control Law. |
| | 7 | Environment Agency established. |
| | 9 | Central Council for Pollution Control established. |
| 1972 | | Urban Parks Improvement (Emergency Measures) Act. |
| | | Industrial Relocation Promotion Act attempts to divert industries from cities. |
| | | Law of Special Measures for Comprehensive Development of Lake Biwa. |
| | 6 | Cabinet approves 'On Environmental Conservation |

| Year | Month | |
|------|-------|---|
| *(1972)* | | Measures Related to Public Works'. |
| | 6 | Nature Conservation Law. |
| | 6 | Air and Water Pollution Control Laws revised for non-fault liability. |
| | 6 | Law Relating to the Regulation of Transfer of Special Birds. |
| 1972–3 | | 'Big Four' lawsuits decided in favour of plaintiffs. |
| 1973 | | City Planning Act revised to include fiscal measures; originally inadequate for discouraging land speculation. |
| | | MITI sets up Industrial Location and Environmental Protection Bureau. |
| | 4 | Nature Conservation Council set up. |
| | 5 | EQS for $NO_2$ and oxidants. |
| | 8 | Emission standards for $NOx$ in stationary sources. |
| | 8 | Interim Law for Conservation of the Environment of the Seto Inland Sea. |
| | 9 | Natural Parks Law and Nature Conservation Law made more stringent. |
| | 10 | Pollution-Related Health Damage Compensation Law replaces Special Relief Law of 1969. |
| | 10 | Cabinet adopts Basic Policy on Nature Conservation. |
| | 10 | Chemical Substances Control Law. |
| | 10 | Urban Greenery Conservation Law. |
| | 10 | Factory Location Law. |
| | 10 | Port and Harbour Law and Public Water Reclamation Law amended to include EIA provisions. |
| | 12 | EQS for aircraft noise. |
| 1974 | | National Land Use Planning Law. |
| | | Reserved Agricultural Areas Act. |
| | 6 | Air Pollution Control Law strengthened to include total emission controls for $SO_2$. |
| | 9 | EQS and effluent standards for mercury strengthened. |
| 1975 | 2 | EQS and effluent standards for PCB. |
| | 7 | EQS for Shinkansen noise. |
| 1976 | 6 | Vibration Regulation Law. |
| 1977 | | OECD report notes need for quality of life improvements in Japan. |
| | | Third National Comprehensive Development Plan includes environmental protection and quality of life considerations, proposes revival of small- and mid-size towns, stresses local autonomy. |

| Year | Month | |
|------|-------|---|
| (1977) | | Water Pollution Control Law revised to include COD standards. |
| | 5 | EA sets up Long-Term Plan for Environmental Conservation |
| 1978 | 4 | Cabinet adopts Basic Plan for Conservation of the Environment of the Seto Inland Sea. |
| | 6 | Wildlife Protection and Hunting Law strengthened. |
| | 6 | Water Pollution Control Law revised to include total pollutant load control. |
| | 7 | EQS for $NO_2$ revised. |
| 1979 | | Law for Promotion of Alternative Energy Supply. |
| | 6 | Total emission control programme for Seto Inland Sea, Tokyo Bay and Ise Bay. |
| 1979–81 | | Ordinances for prevention of eutrophication of lake water. |
| 1980 | | Law of Environmental Development along Trunk Roadside Zones. |
| | | District Planning Act. |
| 1981 | | EQS for nitrogen and phosphorous in lakes. |
| | 4 | Draft EIA Law presented to Diet. |
| | 6 | Air Pollution Law amended to include total emission control for $NOx$. |
| 1982 | 12 | EQS for nitrogen and phosphorus in lakes and reservoirs revised. |
| 1983 | | 'Technopolis Law'. |
| 1984 | 7 | Law Concerning Special Measures for Conservation of Lake Water Quality. |
| | 8 | Cabinet Decision on EIA. |
| | | Agriculture Promotion Areas Act revised to require employment and recreation to be considered. |
| | | National Land Use Plan revised to encompass urban, agricultural, forest, natural park and nature conservation areas. |
| 1985 | 5 | EQS for nitrogen and phosphorus set. |
| 1987 | 4 | Total Water Pollutant Control Programme for Tokyo Bay, Osaka Bay and Seto Inland Sea. |
| | 6 | Fourth National Comprehensive Development Plan. |
| | 11 | Pollution-Related Health Damages Compensation amended to terminate designation of class I. |
| 1988 | 5 | Law to Protect the Ozone Layer by Regulating Certain Substances. |
| | 6 | Provisional guidelines on tree planting for air purification. |

| Year | Month | |
|------|-------|---|
| *(1988)* | 9 | Japan ratifies the Montreal Protocol on Protection of the Ozone layer. |
| | 10 | EA changes regulations controlling use of pesticides. |
| | 12 | EA changes maximum permissible levels for car exhaust gases. |
| 1989 | 6 | Partial changes to Air Pollution Control Law and Water Pollution Control Law. |
| | 7 | EA changes regulations on limits for nitrogen and phosphorous content of drainage water entering lakes. |
| | 8 | Partial amendment to the marine Pollution Prevention Law. |
| | 9 | Amendments to the Offensive Odours Control Law. |
| 1990 | 3 | Amendment to the Pollution-Related Health Damage Compensation Law increases compensation to victim's families. |

Source: Industrial Pollution Control Association of Japan (1983), p. 166; Environment Agency (various years), *Quality of the Environment in Japan*; Keizai Shunjusha Co. (1972), pp. 10–19; Ministry of Foreign Affairs (1974), pp. 11–13.

# Laws by topic

| Topic | Area of responsibility | Relevant laws | Date enacted/ amended | Jurisdiction |
|---|---|---|---|---|
| [1] General pollution control measures | Planning of pollution control measures | Basic Law for Environmental Pollution Control | Aug. 1967/1970, 1971 | EA |
| | Establishment of environmental quality standards | | | EA |
| | Formulation of environmental pollution control programmes | | | EA, EPA, MITI, etc. |
| [2] Air pollution | Establishment and enforcement of standards for emissions from factories | Air Pollution Control Law | June 1968/1970, 1971, 1972 | EA |
| | | Electric Power Industry Law | July 1964 | MITI |
| | | Gas Industry Law Mine Safety Law | | MITI |
| | Establishment and enforcement of automobile emission standards | Air Pollution Control Law Road Transportation Law | As above | MoT, EA |
| | | Road Traffic Law | | Police Agency |
| | Measures for the control of smoke and soot emitted from household heating systems etc. | Air Pollution Control Law | As above | EA |
| [3] Water pollution | Establishment and enforcement of standards for effluent from factories | Water Pollution Control Law | Dec. 1970/ 1971, 1972 | EA |
| | | Electric Power Industry Law | July 1964 | MITI |
| | | Mine Safety Law | | MITI |
| | Control of water pollution caused by effluent from sewerage systems | Sewerage Law | April 1958/1970 | EA, MoC |
| | Control of marine pollution caused by wastes from vessels | Marine Pollution Prevention Law | Dec. 1970/1989 | EA, MoT |
| | Water contamination control | Water Contamination Prevention Law | Dec. 1970 | |
| | Control relating to rivers | River Law | July 1964 | MoC |
| | Control relating to lakes | Clean Lakes Law | July 1984 | EA |
| | Conservation of fishery resources | Fishery Resource Conservation Law | | MAFF |
| | Establishment of potable water quality standards | Water Works Law | | MHW |

| Topic | Area of responsibility | Relevant laws | Date enacted/ amended | Jurisdiction |
|---|---|---|---|---|
| | Area specific controls | Seto Inland Sea Environmental Preservation Law | Aug. 1973/1978 | EA |
| [4] Noise and vibration | Establishment and enforcement of factory noise standards | Noise Regulation Law | June 1968/1970, 1971 | EA |
| | | Electric Power Industry Law Gas Industry Law Mine Safety Law | July 1964 | MITI |
| | Establishment and enforcement of standards for noise emanating from construction sites | Noise Regulation Law | As above | EA |
| | Establishment and enforcement of automobile noise standards | Noise Regulation Law Road Transportation Law Road Traffic Law | As above | As above |
| | Measures for the control of aircraft noise | Law concerning Prevention, etc., of Disturbance Caused by Aircraft Noise in the Vicinity of Public Airports | Aug. 1967 | MoT |
| | | Adjustment, etc., in the Environs of Defence Facilities Law | | Defence Agency |
| | Control of other kinds of noise | Minor Offence Law | | Police Agency |
| | Control of vibration | Vibration Regulation Law | June 1976 | |
| [5] Ground subsidence | Basic measures to prevent ground subsidence | | | EA |
| | Control of pumping of groundwater for industrial use | Industrial Water Law | June 1956/1962, 1964, 1966, 1971, 1972 | MITI |
| | Control of pumping of groundwater for use in buildings | Law Concerning Regulation of Pumping of Ground Water for Use in Buildings | May 1962/1964, 1971 | EA |
| | Measures to prevent subsidence of agricultural land | | | MAFF |
| [6] Offensive odours | Control of offensive odours emanating from plants, etc., processing dead animals | Law Relating to Dead Animal Processing Plants, etc. | | |
| | Offensive odour control | Offensive Odour Control Law | June 1971/1989 | EA |

| Topic | Area of responsibility | Relevant laws | Date enacted/ amended | Jurisdiction |
|---|---|---|---|---|
| [7] Soil pollution | Soil pollution control and measures for cleansing polluted soil | Agricultural Soil Pollution Prevention, etc., Law | Dec. 1970/1971 | MAFF |
| [8] Waste disposal | Disposal of industrial and non-industrial wastes | Waste Disposal and Public Cleansing Law | Dec. 1970 | MHW, MITI, EA |
| | Poisonous and deleterious substances control | Poisonous and Deleterious Substances Control Law | Dec. 1950/1970 | MHW |
| [9] Agricultural chemicals | Establishment of standards and registration system for agricultural chemicals | Agricultural Chemicals Control Law | July 1948/1971 | MAFF, EA |
| [10] Control of land utilization and construction of facilities | City planning | City Planning Law Building Standards Law | June 1968 | MoC MoC |
| | Control of new and/or additional construction of factories | Law Concerning Restriction on Industries, etc., in Built-up Districts in the National Capital Region | March 1959 | Commission for the Development of the National Capital Region |
| | | Law Concerning Restriction on Industries, etc., in Built-up Districts in the Kinki Region | July 1964 | Commission for the Development of the Kinki Region |
| | Survey of conditions affecting the location of plants | Factory Location Law | Oct. 1973 | MITI |
| | Control of reclamation | Public Water Areas Reclamation Law | April 1921/1973 | MoC, MoT, MAFF, EA |
| [11] Improvement of pollution control facilities and conservation of nature | Regional development planning | National Capital Region Development Law | 1956 | NLA |
| | | Law for the Conservation of Green Belts around the National Capital Region | June 1966 | NLA |
| | | Kinki Region Development Law | 1963 | NLA |
| | | Law for the Development of Conservation Areas in the Kinki Region | July 1967 | NLA |
| | | Chubu Region Development Law | July 1967 | NLA |

| Topic | Area of responsibility | Relevant laws | Date enacted/ amended | Jurisdiction |
|---|---|---|---|---|
| | Development of new industrial cities and industrial development of special areas | Law for Promoting Development of Special Areas for Industrial Consolidation | 1964 | EPA, MAFF, MITI, MoT, MoC, MHA |
| | | Law for Promoting the Establishment of the New Industrial Cities | 1962 | |
| | | Factory Location Law | Oct. 1973 | MITI |
| | Agricultural development | Law for Improvement of Agricultural Promotion Areas Agricultural Land Law | | MAFF |
| | Construction of sewerage systems | Sewerage Law | April 1958/1970 | MoC |
| | | Law Concerning Emergency Measures for Sewerage Construction | | |
| | Construction of buffer zones | City Planning Law | 1968 | MoC |
| | City park develop- ment | City Parks Law | April 1956 | MoC |
| | | City Green Zone Conservation Law | | MoC |
| | Conservation of the environment under the natural parks system | Natural Parks Law | June 1957/1962, 1970, 1971, 1972 | EA |
| | Nature conservation | Nature Conservation Law | June 1972 | EA |
| | | National Land Use Planning Law | 1974 | NLA |
| | Forestry conserva- tion | Forestry Law | June 1951 | MAFF |
| | Coastal conserva- tion | Coastal Law | | MoT, MoC, MAFF |
| | Protection of wildlife | Law Concerning Wildlife Protection and Hunting | April 1918/1970, 1971, 1972 | MAFF |
| | Protection of trees | Law Concerning the Protection of Trees for the Conservation of Scenic Beauty | May 1962 | MoC |
| | Protection of cultural properties | Law Concerning Special Cultural Properties | May 1950 | Culture Agency |
| | | Law Concerning Special Measures | Jan. 1966 | PM's office, MoC |

| Topic | Area of responsibility | Relevant laws | Date enacted/ amended | Jurisdiction |
|---|---|---|---|---|
| | | for Preservation of Historical Natural Features of Ancient Cities | | |
| [12] Settlement of disputes and relief | Settlement of environmental pollution disputes | Pollution Disputes Settlement Law | June 1970 | PM's office |
| | | Mining Law | | |
| | | Temporary Law for Compensation of Damage Caused by Coal Mines | | MITI |
| | | Pollution-Related Health Damage Compensation Law | Oct. 1973/1987 | EA |
| | Relief for patients affected by environmental pollution | Special Measures for the Relief of Pollution Related Patients Law | Dec. 1969/1971/ 1990 | EA, MITI, PM's office |
| [13] Cost bearing and incentive measures | Determination of entrepreneurs' share of the cost of public pollution control works | Law Concerning Entrepreneurs Bearing the Cost of Public Pollution Control Works | Dec. 1970 Dec. 1970 | PM's office |
| | Loans from the Environmental Pollution Control Service Corporation | Pollution Control Services Corp. Law | June 1965/ 1968, 1971 | MITI, MoC, MAFF |
| | Loans for modernization of small and medium enterprises | Law for Loans for the Modernization of Small and Medium Enterprises | | |
| | Pollution control projects | Special Government Measures for Pollution Control Projects | May 1971 | |
| | Special taxation measures | Corporate Tax Law | | Min. of Finance, MHA |
| | | Special Taxation Measures Law | 1957 | |
| | | Local Tax Law | | |
| | Subsidies for noise control measures in vicinity of public airports | Law Concerning Prevention, etc., of Disturbance Caused by Aircraft Noise in the Vicinity of Public Airports | Aug. 1967 | MoT |
| | Subsidies for noise control measures in areas surrounding defence facilities | Law Concerning Adjustment, etc., in the Environs of Defence Facilities | | Defence Agency |

| Topic | Area of responsibility | Relevant laws | Date enacted/ amended | Jurisdiction |
|-------|------------------------|---------------|----------------------|--------------|
| | Subsidies for pollution control facilities of schools | | | Ministry of Education |
| [14] Punishment of crimes relating to environmental pollution | Punishment of crimes and offences relating to environmental pollution | Law for the Punishment of Environmental Pollution Crimes relating to Human Health | Dec. 1970 | Ministry of Justice |
| | Protection of human rights | | | Ministry of Justice |
| [15] Other | Regulation of toxic substances, etc., used in factories | Labour Standards Law | | Ministry of Labour |
| | Chemical substances | Chemical Substance Control Law | Oct. 1973/1979, 1986 | MHW, MITI |
| | Toxic substances | Toxic Substances Law | Dec. 1950/1970 | MITI |
| | Radioactive substances | | | |

# Reports related to the implementation of Japan's national EIA system

**Environment Agency, Central Council on Environmental Pollution Control (1974), 'Guidelines for conducting environmental impact assessment (interim report)'**

The concept of environmental impact assessment – foretelling and assessing the impact of a development project on the environment before it gets under way and limiting development to the extent that preservation of the environment is ensured – is being gradually accepted in administration. It has resulted in such measures as revision of the standards on issuance of permits and licensing under the relevant statutes and expansion of matters requiring consultation with the Director-General of the Environment Agency.

Against this background, the environmental impact assessment sub-committee of the Central Council for Control of Environmental Pollution has presented an interim report of the results of its deliberations on guidelines to be followed in carrying out environmental impact assessments. The proposed guidelines, which particularly bear on large-scale industrial development projects, are outlined below.

## 1. Significance of environmental impact assessment

It is essential that the environmental impact of a development project should be sufficiently assessed and anticipated prior to its execution. The project must not be started until environmental conservation is guaranteed through the impact assessment process. The objective of this operation is to foresee and assess (or reassess) beforehand the dimensions of the effects the planned development would have on the environment, such as the quality of air, water and soil and living organisms. It includes a search for measures that would prevent adverse effects and a comparative study of alternative plans.

The idea of comparing merits and demerits of development has not been fully incorporated in environmental assessment hitherto carried out in Japan. Environmental attention in this country has centred on how to cope with the more immediate problems of health and other pollution damages, and the comparison of human life and other absolute values with economic values was regarded contrary to the popular sentiment.

Such being the case, it is of vital importance to set appropriate environmental protection standards and to make objective appraisal of the environmental impact of future development in relation to these standards.

The implementation of environmental impact assessment could entail the following basic questions:

### (1) Timing of impact assessment

An assessment needs to be carried out in each stage of planning. The more immature planning is, the more uncertain what changes the projected development would bring about and the less precise its impact assessment, making it necessary to allow for the greater margins of safety.

### (2) Scientific limitations of impact assessment

An environmental assessment must be backed up by scientific knowledge, but its accuracy is necessarily circumscribed to the degree that the scientific knowledge involved is only the best available at the time. This gives rise to the need to make a distinction between what is scientifically certain and what is not and to identify clearly the hypothetical conditions.

### (3) Importance of reassessment

The fact that an environmental impact assessment has to be performed by using hypotheses and with the best scientific knowledge available at the time gives particular importance to constant surveillance in the subsequent period and follow-up checks, with feedbacks to be made on the basis of newly acquired information. If a reassessment shows that there is a threat to environmental conservation needs, a review of the development projects in question must immediately be started.

## 2. Impact assessment and environmental quality standards

### (1) Standards relating to human health and living environment

With regard to air, water and soil pollution and noise, the existing environmental quality standards 'desirably to be observed to protect human health and preserve the living environment' can, in principle, also properly serve as environmental conservation standards to be adhered to in performing environmental impact assessments.

Some action is also in order for living environment items not covered by the environmental quality standards, such as working out provisional protection standards if a sufficient amount of scientific knowledge is amassed to do so.

For example, some kind of consideration is warranted to deal with poisonous substances that cause trouble between residents and local factories, the effects of heated water discharged by thermal power stations, and the excessive presence of nitrogen and phosphorous in enclosed bodies of water.

As for the standard for offensive odour, it should be 'a level at which the majority of local inhabitants do not become aware of it in their daily lives'. For ground subsidence, it should be 'a rule that will not allow further subsidence'.

### (2) Standards relating to natural environment

Because it is impossible to work out all-embracing absolute standards for the preservation of the natural environment, four classes of valuable nature – valuable for scientific interests, scenic beauty and outdoor recreational use – should be established on the basis of rarity, distinctiveness and uniqueness with a different conservation standard to be set for each division. The four ranks are according to degrees of value – the natural environment of national value, of regional value, of prefectural value, and of value on a city–town–village level.

## 3. Gathering of information and environmental forecasting

### (1) Gathering of basic information

The most essential prerequisite to the forecasting of the impact of development on the environment is to gather basic information on meteorological, hydrological, geographical and other natural conditions. Since it is practically impossible to collect this necessary information in a short period of time, adequate attention must be paid to the matter from the initial stage of planning.

### (2) Factors and methods

*Air pollution* In addition to sulphur dioxide, suspended particulate matter, nitrogen oxides and carbon monoxide, predictions are to be made, where necessary, on the spread of harmful substances which are expected to be discharged by planned facilities. The standard procedure would be to use a numerical calculation formula based on a dispersion theory.

*Water pollution* Besides BOD, COD, SS and oil, forecasts are to be made, where necessary, on the dispersion of thermal water from power stations and harmful substances, obtaining sufficient information on changes in currents in the case of lakes, marshes and sea areas. A numerical calculation formula based on a dilution theory is to be relied on in principle. The Helps formula is to be used for rivers, taking account of a future decline in the flow of water as the current flows in one direction only.

*Effects on natural environment* Predictions on the effects of development on the natural environment, mainly with regard to qualitative changes, should cover the felling of trees, excavations, land reclamations, land levelling for industrial or housing use, engineering works, and changes in water levels and water temperatures.

*Others* Forecasts should also be carried out on required items regarding ground subsidence, noise, vibration, odour, and wastes.

**Environment Agency, Central Council on Environmental Pollution Control (1975) 'Legal system of environmental impact assessment'**

*1. Basic character of EIA system*

The environmental impact of a development project is to be assessed by the executor before the project gets under way and on the basis of adequate research. The results of appraisal are to be made public, and opinions are to be solicited from specialists who possess relevant scientific knowledge and local inhabitants, so that what these people say can be reflected in the final evaluation. The proposed legal system is to be basically concerned with these procedures.

*2. Extent of system's applicability*

(1) The projected system basically is to apply to development plans which could have a marked effect on the environment. A decision on whether a specific development project comes under the system must be based on a statute and the like. This makes it necessary for the national government to require environmental impact assessment reports before authorizing non-state projects, such as the construction of a factory complex, a dam, a facility for high-speed transport and so on.

In the case of a massive regional development plan with multiple elements, an environmental impact assessment of the entire project is desirable before individual component plans are brought under the same scrutiny. A search for ways to ensure such a process by designating such projects as 'large-scale plans' is in order.

(2) Development plans normally take shape as they go through a series of decision-making stages. An environmental evaluation under the

proposed system is to be carried out when a plan has reached a certain stage of maturity.

(3) It goes without saying that even in cases which fall outside the province of the EIA system, appropriate attention is to be paid to environmental preservation when an environmental impact is involved.

## 3. Basic principles of assessment

(1) What environmental impact assessment should involve is determining in advance the degree of a development plan's impact on environmental resources, such as air, water, soil and living creatures and so on, and figuring out ways to prevent expected environmental disruptions, including a comparative study of alternative plans.

(2) In concrete terms, what is to be done is to estimate in advance changes in the quality of the environment which would result from the execution of a development project and to determine whether and how much these changes would make the environment better or worse than the level of quality deemed necessary to keep human beings healthy and fulfil other environmental requirements. Thus, the first thing to do is to establish the essential level of environmental quality.

(3) Some standards need to be developed by which it is possible to scientifically judge whether a particular level of environmental quality can be considered acceptable.

Under the EIA system, the first step should be to work out a preliminary version of the essential level of environmental quality in each case by taking account of what experts suggest, the needs of the local community, and the general desires of inhabitants. The preliminary decision then should be made public so that all local residents have a chance to express their opinions. This procedure will make the final decision more objective.

(4) In performing an assessment, consideration of alternative plans is desirable. The circumstances in Japan make it difficult to prepare plans of markedly different nature or those which constitute a drastic revision of the original. But steps like mere clarification of the process by which the original plan has been put into shape would be similarly beneficial.

(5) Items to be involved in an environmental appraisal have to include all conceivable changes in the quality of the environment that may result from the execution of the development project in question. Steps must be taken to make sure that the evaluation is made in a proper manner, such as laying down a guideline on items to be checked on and techniques to be used.

### 4. Importance of reflecting inhabitant opinions

(1) Soliciting opinions from the public is to be an important procedure under the proposed EIA system. Reflecting the desires of local inhabitants and conservationist groups in development plans is a procedure that has not been generally practised, but is to be a cornerstone of the system.

(2) In institutionalizing environmental impact assessment, it is indispensable to know what local inhabitants want of the quality of the environment, apart from human health and other items about which a general consensus can be easily obtained.

Experience-based points made by local inhabitants and recommendations by experts are precious information which is hard to obtain unless the EIA system is instituted.

(3) The system is to be so designed that there will be adequate exchanges of information between the executor of the development project and local inhabitants as to the possible environmental effects of the plan, mutual understanding will grow between the two parties, effective deliberations will be made from the viewpoint of environmental preservation, and the desires of local inhabitants will be respected.

The objective in soliciting local opinions should be not to find out whether the development plan is supported by a majority of the local populace but to reflect them in the process of environmental impact assessment.

### 5. Main EIA procedures

The basic procedures to be followed under the proposed EIA system are:

(1) The would-be executor of a development project estimates in advance the environmental impact of the plan based on adequate research on the environment impact assessment report.

(2) The findings are made public with steps to be taken to acquaint the public with them, such as the holding of gatherings at which they are explained.

(3) Local inhabitants are invited to express their opinions in writing, at public hearings or in other manners.

(4) Opinions are solicited from the Director-General of the Environment Agency and chiefs of local governments concerned.

(5) When these steps are completed, the would-be executor of the plan reviews the preliminary findings of assessment on the basis of expressed opinions and making amendments where necessary, prepares the final environmental impact assessment report which lists what the executor thinks of the voiced views and measures to deal with them.

(6) In the case of a development plan requiring state approval or

authorization, the impact assessment is to be made by the applicant for permission, while the procedures under the second and fifth headings are taken care of by authorities with the power to grant clearance (in the latter case, what authorities should do is to draw up documents on views expressed about the preliminary environmental impact assessment report).

**Environment Agency, Central Council on Environmental Pollution Control (1979), 'Recommendation on a system of environmental impact assessment'**

*1. Method of institutionalization*

Practical experiences are being gained in Japan, while efforts are also being made in other developed nations to set up a system of environmental impact assessment suited to the varied legislative, institutional and administrative frameworks. In Japan, at present, environmental impact assessment is made on different types of projects at the national level in line with specific laws under the jurisdiction of the pertinent ministries and agencies, or through their administrative guidance and other administrative actions. At the local government level, environmental impact assessment is carried out in accordance with ordinance or enforcement manuals. However, these activities vary in their purpose and object of assessment, the timing of survey, forecasting and assessment, the scope of environmental preservation and the procedures involved. It is therefore essential to establish general rules for an environmental impact assessment system as soon as possible. Although various methods are conceivable, the enactment of law would best meet the end.

*2. Purpose and basic framework of environmental impact assessment system*

The prime objective of the environmental impact assessment system is to integrate environmental consideration into the planning and decision-making process of a project along with social and economic factors. The second objective is to clarify the respective roles of related administrative authorities and the form of public investment by residents of the localities concerned, and to incorporate their comments into the planning and decision-making process of a project. These steps will facilitate understanding by the residents of the localities concerned and thereby serve a proper implementation of the project.

Accordingly, the environmental impact assessment system should mainly be concerned with the design of procedures whereby a project agency or an entrepreneur shall survey, forecast and assess potentially significant environmental impacts, and invite comments concerning environmental preservation from administrative authorities and from the residents of the localities concerned on the basis of the results of such assessment.

### 3. Projects to be covered by the system

The system should cover projects of a considerably large scale which could have significant impact on the environment. The scope of such projects should be specifically determined in advance. As to regional development plans, port and harbour plans should also be covered by the system. As regards plans related to development projects for specially designated areas, environmental impact assessment should be carried out in each case.

### 4. Timing of environmental impact assessment

Environmental impact assessment should be carried out at such a time as will make it easier to have measures for environmental preservation incorporated into the project and as the project is substantiated so that the assessment can be implemented in an effective manner. Moreover, it is necessary to determine in advance the time limit for circulation of, and receiving comment of, the draft impact statement.

### 5. Alternatives

In Japan, a densely populated island country, possible alternatives are limited, especially in regard to the selection of a site for a project. When the choice of alternatives is made prior to the preparation of draft environmental impact statement, it will be appropriate to indicate in the statement the alternatives already studied as measures for prevention of environmental pollution or conservation of the natural environment.

### 6. Public involvement

Involvement of residents of a locality concerned is designed not so much to measure the degree of approval of or opposition to the project itself, as to have their comments concerning environmental preservation incorporated into the planning and decision-making process of the project. It will be appropriate to limit the scope of residents who can submit comments to the project agency or the entrepreneur to people living in the locality where the project is to be carried out and where the environment might be affected as a result of implementation of the project.

In order to make the residents acquainted with the project, the project agency or the entrepreneur should take such steps as circulation of a draft environmental impact statement and holding of meetings.

### 7. Roles of local and national governments

The local governments concerned, the competent ministers and Director-General of the Environment Agency should endeavour to complement and improve the contents of a survey, forecast and assessment carried out by the project agency or the entrepreneur by such means as submission of relevant data and expression of their comments.

### 8. Ensuring effectiveness

It would be essential to withhold at least the start-up of the project until environmental impact assessment is completed, so that the results of the assessment may be incorporated into the planning and decision-making process of the project.

### Cabinet decision (28 Aug. 1984), 'On implementation of environmental impact assessment'

(1) Considering that carrying out environmental impact assessment before projects are undertaken is of critical importance for pollution control and for conservation of the natural environment, the government establishes the following scheme for the implementation of environmental impact assessment, specifying procedures and others to be followed.

(2) In order to implement environmental impact assessment, national administrative agencies should, as expeditiously as possible, take necessary steps based on this scheme for the projects implemented with their licences, etc.[1]

(3) The government should seek the understanding and cooperation of project undertakers and local governments to ensure that steps based on this scheme will be effectuated smoothly.

(4) The government should ask local governments to respect the spirit of this Cabinet decision and to pay due attention to conformity with this scheme when they take measures concerning environmental impact assessment.

(5) As stipulated in the Appendix, the Committee for Fostering the Implementation of Environmental Impact Assessment should be established under the Cabinet in order to decide details commonly needed for the effectuation of the procedures and others prescribed in this scheme, including those which this scheme requires to be decided separately.

## Implementation scheme for environmental impact assessment

### 1. Relevant project and others

(1) Of the following categories of projects, relevant projects subject to this scheme should be designated by the competent ministers in consultations with the Director-General of the Environment Agency as large-scale projects the implementation of which is likely to cause significant environmental impacts – those related to environmental pollutions (excluding those caused by radioactive substances) and the natural environment:

(a) Construction or reconstruction of national expressways, national roads, and other roads.

(b) Construction of dams on rivers prescribed by the River Law and river waterworks prescribed by the same law.

(c) Construction or improvement of railways.

(d) Construction of airports and changes of their facilities.

(e) Reclamations and dumping works.

(f) Land readjustment projects prescribed by the Land Readjustment Law.

(g) New residential built up area development projects prescribed by the New Residential Built Up Area Development Law.

(h) Industrial estate construction projects prescribed by the Act concerning the Development of the Suburban Development and Redevelopment Area and Urban Development Areas in the National Capital Region and those prescribed by the Act concerning the Development of Suburban Development and Redevelopment Areas and Urban Development Areas in the Kinki Region.

(i) New urban infrastructure development projects prescribed by the Law for Development of the Basis of New Cities.

(j) Distribution business centre construction projects prescribed by the Law concerning Construction of Distribution Business Centres.

(k) Projects implemented by corporations established by special legislation for preparing land for residential purposes, sites for factories and other business establishments, and land for other purposes.

(l) Any other projects similar to the above designated by the competent ministers in consultation with the Director-General of the Environment Agency.

(2) Environmental impact assessment should be implemented by project undertakers who implement relevant projects and who would be designated separately.

## 2. Procedures and others for environmental impact assessment

(1) Preparation of draft environmental impact statement (EIS)

(a) In implementing the relevant projects, the project undertakers should make surveys and studies, prediction and evaluation of the environmental impacts of such projects. [In the case of projects other than those coming under the category 1–(1)e, such impacts should include those caused by business and other human activities which will be carried out on the land (exemptible if the land is subject to the use for other relevant projects) or at the facilities after completion of the projects, and should exclude the impact of reclamations and dumping works carried out for the relevant projects].

The project undertakers should prepare a draft EIS, covering the following items:

(i)   Name and address of the project undertaker.
(ii)  Purpose and description of the project.
(iii) A summary of findings of surveys and studies.
(iv)  Statement of environmental impact of the project and its magnitude and measures planned for pollution control and for conservation of the natural environment.
(v)   Evaluation of environmental impact of the project.

(b) Surveys and studies, prediction and evaluation mentioned in (a) above should be conducted in accordance with guidelines which should be established for each category of relevant projects by the competent minister in consultations with the Director-General of the Environment Agency. In consultations with the heads of related administrative agencies, the Director-General of the Environment Agency should formulate the basic principles to be taken into consideration by the competent ministers when they establish the guidelines.

(2) Publicity concerning draft EIS

(a) The project undertaker should send the draft EIS to the prefectural governor and the mayors with jurisdiction over the related area. In cooperation with them, the project undertaker should announce publicly that the draft EIS has been prepared. The draft EIS should then be made available for public review for a period of one month from the day of such announcement.

(b) During the period when the draft EIS is available for public review, the project undertaker should hold explanatory meetings on the project in the related area. When the explanatory meetings cannot be held for reasons not attributable to the responsibility of the project undertaker, it is not necessary to hold the meetings. In such a case, however, the project undertaker should make efforts to publicize the draft EIS by other means.

(3) Comments on draft EIS

(a) The project undertaker should invite the comments of the residents of the related area on the draft EIS from the perspective of pollution control and conservation of the natural environment (comments are valid only when they are submitted in writing during the period of the public review and the following two weeks).

(b) The project undertaker should send a report summarizing the comments of (a) above to the prefectural governor and the mayors with jurisdiction over the related areas, and ask the prefectural governor to make, within three months after receiving the summary, comments on the draft EIS from the perspective of pollution control and conservation of the natural environment after soliciting the views of the mayors.

(4) Preparation of final EIS

(a) After the comments on the draft EIS are received or after the period specified in (3)(b) expires, the project undertaker, by reviewing the draft EIS, should prepare a final EIS on environmental impact assessment covering the following items:

(i)   Items listed in 2.(1)(a)(i)–(v)
(ii)  A summary of the comments received from the residents of the related area
(iii) Comments of the prefectural governor with jurisdiction over the related area
(iv)  Views of the project undertaker on the comments received from the residents and prefectural governor

(b) The project undertaker should send a final EIS to the prefectural governor and the mayors with jurisdiction over the related area, and with their cooperation, he should announce publicly that the final EIS has been prepared. The final EIS should be made available for public review for a period of one month from the day of the announcement.

(5) Other procedural matters

(a) The project undertaker, after consultations with prefectural and other local governments, can entrust the holding of explanatory meetings, etc. to these local governments.

(b) In cases in which local governments implement projects with subsidies and other state financial aid, the state should give appropriate consideration to the expenses needed for the implementation of the required procedures and others of environmental impact assessment.

*3. Incorporation in national administration with regard to pollution
control and conservation of the natural environment*

(1) Sending final EIS to national authorities

(a) The project undertaker should send the final EIS promptly after
the public announcement of its completion to relevant authorities
specified separately when the relevant projects are subject to licences,
etc., and to the Director-General of the Environment Agency when the
relevant projects are state ones.

(b) The head of a state administrative agency, after receiving a final
EIS in accordance with (a) above, should send it promptly to the
Director-General of the Environment Agency.

(2) Views of the Director-General of the Environment Agency

If special consideration concerning environmental impact is deemed
necessary because of the large scale of a certain project among those for
which final EISs were sent to the Director-General of the Environment
Agency in accordance with (1) above, the competent minister should
solicit views of the Director-General of the Environment Agency on the
final EIS from the perspective of pollution control and conservation of
the natural environment.

(3) Review of consideration given to pollution control and conservation
of the natural environment

(a) When issuing licences, etc., relevant authorities should review the
contents of the final EIS to ensure that appropriate consideration is
guaranteed for pollution control and conservation of the natural
environment in its implementation, within the jurisdiction prescribed in
the provisions of the laws concerning such relevant licences, etc.

(b) When the Director-General of the Environment Agency expresses
views in accordance with (2) above, such views should be expressed
before the review mentioned in (a) above. The authorities empowered to
issue licences, etc. should carry out the review by paying due attention to
such views within the jurisdiction prescribed in the provisions of the laws
concerning relevant licences, etc.

(c) The project undertakers should give consideration to the environmental impact of the relevant project set out in the final EIS. When the
views of the Director-General of the Environment Agency have been
expressed in accordance with (2) above, the project undertakers should
pay due attention to these views and should implement the projects with
appropriate consideration to pollution control and conservation of the
natural environment.

### 4. Other matters

(1) The details needed for the effectuation of the procedures and others
prescribed in this scheme, including those to be decided by the competent ministers and those to be decided separately, should be decided as expeditiously as possible. However, the basic principles specified in 2(1)(b) above and other details commonly needed for the effectuation of the procedures and others prescribed in this scheme should be determined within three months from the day of this Cabinet decision.

(2) Provisional arrangements concerning the implementation of this scheme should be specified separately.

### Environment Agency (27 Nov. 1984), 'Principles concerning surveys and studies, prediction and evaluation of environmental impacts'

### 1. Objective

The provisions hereafter specify the principles for surveys and studies, prediction and evaluation (hereinafter referred to as surveys and the like) of environmental impacts which shall be taken into consideration by the competent minister in establishing guidelines under 2(1)b of the Scheme for the Implementation of Environmental Impact Assessment.

### 2. General principles

(1) The surveys and the like of the environmental impact of the project shall be carried out in accordance with guidelines which are established by the competent minister for each category of project specified in the scheme.

(2) The surveys and the like shall be carried out concerning pollution control and conservation of natural environment as follows.

(a) The surveys and the like concerning pollution control shall cover the items relating to the protection of human health and the conservation of the living environment including property closely related to human life and the animals and plants related to human life and their habitat.

(b) The surveys and the like concerning the conservation of natural environment shall cover the items relating to the proper conservation of various categories of nature such as wilderness areas, irreplaceable nature which has great scientific and cultural value, outstanding natural scenery, the habitat of wild animals, and natural areas fit for outdoor recreation activities.

(3) The surveys and the like shall be conducted about the following actions for execution of the project and according to the project at the time of surveys.

(a) With respect to the projects other than land fill and reclamation:

(i) Works for the execution of the project (except land fill/reclamation works for the execution).

(ii) Existence of the land (except that for other projects) or the structures completed by the works specified in (i).

(iii) Business and other human activities planned to be conducted on the land or at the structure specified in (ii).

(b) With respect to the project of land fill or reclamation:

(i) Works for the execution of the project.

(ii) Existence of the land (including banks, wharves and other similar structures) completed by the works specified in (i).

(4) The guidelines shall specify, based on the available scientific knowledge, the items of surveys and the like, which are generally deemed necessary to identify the environmental impact of the project, and the reasonable technical methods to be used for the surveys and the like to identify such impact.

(5) The guidelines shall be established so as to ensure the propriety of the surveys and the like by paying attention to the characteristics of the project and those of the area to be surveyed (hereinafter referred to as the survey area) on the environmental impact of the project.

## 3. Items of survey and the like and their handling

(1) The items of surveys and the like, which are generally deemed necessary to identify the environmental impact of the project (hereinafter referred to as items to be covered) shall be specified in the guidelines as necessary items in accordance with the characteristics of each project concerning the elements of the environment listed in the attached table.

*Elements of the Environment*
Elements relating to pollution control
— Air pollution
— Water pollution
— Soil pollution
— Noise
— Vibration
— Ground subsidence
— Offensive odour

Elements relating to conservation of the natural environment
— Topographical and geological features
— Plants

— Animals
— Scenery
— Outdoor recreation fields

(2) The handling of surveys and the like concerning the items to be covered shall be pursuant to the following viewpoints.

(a) With reference to the items concerning pollution control, it is a principle to make quantitative prediction where it is possible and, where it is difficult, to make qualitative prediction, based on the survey findings.

And as for the evaluation, it is a principle to evaluate, taking into consideration the results of surveys, prediction and deliberation about pollution control measures, the environmental impacts in the light of the environmental quality standards or criteria based on scientific knowledge where they are available and, where such evaluation is difficult, through judgement based on scientific knowledge.

(b) With reference to the items concerning the conservation of the natural environment, surveys and analysis are to be done about the existing state of the natural environment in the survey area, identification of the conditions of the essential natural environment in the areas is to be made based on the findings thereof, quantitative or qualitative prediction is to be made as to the changes in the conditions of the essential natural environment, and evaluation is to be made taking into account the level of conservation according to the importance of the natural environment.

Concerning the above, as for the animals among environmental elements, consideration is to be given, where necessary, from the viewpoint of the ecological impact of the project.

### 4. Principles concerning surveys

(1) The surveys shall be carried out by collection of information necessary for prediction and evaluation of the environmental impact of the project and arrangement and analysis thereof. The technical methods to be used for surveys shall be specified in the guidelines.

(2) The surveys shall comprise the collection of existing data, on-the-spot field survey and so on concerning the following points and the arrangement and analysis of their results. Regarding the items requiring prediction and evaluation, care is to be taken so as to ensure the level necessary to fulfil the ends.

(i) The present state of the environment concerning the items to be covered.

(ii) The meteorological, hydrological and other natural conditions and the demographic, industrial and other social conditions, about which information must be collected in connection with the items provided in (i) and which are specified in the guidelines.

(3) The duration and frequency of the survey concerning the items to be covered or the matters to be considered in setting them shall in principle be specified in the guidelines.

(4) The scope of the survey area concerning the items to be covered shall be in principle the area including the sphere where the state of the environment will be changed to a certain degree as a result of the project, or the area where the environment will be transformed directly and its periphery. The scope is to be specified in advance where it is possible, or in other cases, when conducting the survey of each project based on the supposition as to degree of environmental impact caused by the project. These formulas shall be provided in the guidelines.

(5) The survey techniques or measuring methods concerning the items to be covered shall be specified in the guidelines. Where they are specified in the notification of the environmental standards or other laws and regulations, they shall be used in principle. In other cases, appropriate techniques or methods shall be specified in the guidelines.

## 5. Principles concerning prediction

(1) The prediction shall be carried out, concerning the items which are deemed necessary as a result of arranging and analysing the survey findings, by identifying the changes in the state of the environment, which will result in ordinary conditions from the execution of the project. The technical methods to be used for prediction shall be specified in the guidelines.

(2) The prediction, regarding the items to be covered concerning pollution control and the items to be covered concerning the conservation of the natural environment, shall be made in the following manner.

The prediction can be made on reference to the measures or policies for pollution control and the conservation of the natural environment implemented by the project undertakers or the government.

(a) With reference to the items to be covered concerning pollution control, the prediction is to be made, taking into account the characteristics of the project, by means of calculations based on simulation models, simulated experiments, references to previous cases or their analysis and so on.

In choosing the methods for making prediction, it is required to pay attention to the characteristics of respective methods, the conditions for their applying, and the features of the survey area.

(b) With reference to the items to be covered concerning the conservation of the natural environment, the prediction as for direct impact is to be made as quantitatively as possible, in accordance with the characteristics of each item, to size up their disappearance or existence and the degree of transformation. The prediction as for indirect environmental impact, when it is made upon necessity, shall mainly rely on the qualitative approach.

(3) The time to be predicted shall be specified in the guidelines so as to identify precisely the environmental impact according to the characteristics of the project.

(4) The area concerning which the prediction is to be made (hereinafter referred to as prediction area) shall be specified in the guidelines within the confines of the survey area.

## 6. Principles concerning evaluation

(1) The evaluation shall be made, taking into account the results of the surveys and the prediction concerning the items to be covered and the deliberation about measures for pollution control and the conservation of the natural environment, by clarifying the views of the project undertaker about the impact of the project on the environment of the prediction area based on scientific knowledge. The technical methods to be employed for prediction shall be specified in the guidelines.

(2) With respect to the items to be covered concerning pollution control, the evaluation is to be made to determine, on the basis of scientific knowledge, whether the project will have ill-effects to the protection of human health or the conservation of the environment. Concerning this, it is a principle to evaluate the environmental impact in the light of the environmental quality standard where it is specified in accordance with Article 9 of the Basic Law for Environmental Pollution Control and in the light of criteria of impact to human health or living environment where they are available.

(3) With respect to the items to be covered concerning the conservation of the natural environment, the evaluation is to be made to determine, based on scientific knowledge, whether the impact to the natural environment in the survey area will impede appropriate conservation thereof in accordance with the importance of the natural environment.

(4) In the evaluation, according to the necessity, the future state of the environment in the area concerned brought about by other projects than the project specified, is to be taken into consideration. (Where the state of the environment in the future is difficult to foresee on the basis of data provided by the government or local governments, present state of the environment is to be taken into consideration.)

The measures implemented by the government or local governments for pollution control and the conservation of the natural environment can also be taken into account.

## 7. *Others*

The guidelines shall be under continuous scientific deliberation and revised when necessary.

## Note

1. Licences, approvals, reports with advice or orders of authorities, orders and supervision to public corporations and national subsidies.

# References

Abe, H. and Alden, J.D. (1988) 'Regional development planning in Japan', *Regional Studies* 22(5): 429–38.

Alden, J.D. (1984) 'Metropolitan planning in Japan', *Town Planning Review* 55(1): 55–74.

Anderson, F.R., Mandelker, D.R. and Tarlock, A.D. (1984) *Environmental Protection: Law and Policy*, Little, Brown & Co., Boston.

Barrett, B.F.D. (Jan. 1989) 'Environmental impact assessment and environmental policy in Japan', Kyoto University.

British Embassy, Tokyo (1988) *Japan's National Development Projects.*

Benedict, R. (1986) *The Chrysanthemum and the Sword*, Charles E. Tuttle Co., Tokyo.

Bidwell, R. (May 1985) 'The gap between promise and performance in EIA: is it too great?', presented at the RTPI conference on Environmental Assessment, University of Manchester, Manchester.

Canter, L.W. (1977) *Environmental Impact Assessment*, McGraw-Hill, New York.

Citizens Group to Consider the Airport Problem (1987, in Japanese) *Screaming Sea, Fight for Shiraho*, Ishigaki.

Committee for Fostering the Implementation of Environmental Impact Statements (Nov. 1984) *Common Matters Necessary for the Procedures etc. Based on Implementation Scheme for EIA*, Tokyo.

Committee to Consider the New Ishigaki Airport Problem (Oct. 1986, in Japanese) *Problems with the New Ishigaki Airport Plan*, Okinawa.

Committeee to Study the Socio-Economic Effects of the New Ishigaki Airport (Feb. 1986, in Japanese) *Report on the Socio-Economic Effects of the New Ishigaki Airport*, Tokyo.

Cook, P.L. (1979) 'Costs of environmental impact statements and the benefits they yield in the improvements to projects and opportunities for public involvement', presented at the United Nations Economic Commission for Europe conference in Villach, Austria.

Corwin, T.K. (1980) 'Economics of pollution control in Japan', *Environmental Science and Technology* 14(2): 154–7.

Craig, A.M. (1975) 'Functional and dysfunctional aspects of government bureaucracy', in Vogel, E.F. (ed.) *Modern Japanese Organisation and Decision-Making*, Charles E. Tuttle Co., Tokyo: 3–32.

Donnison, D. and Hoshino, S. (1988) 'Formulating the Japanese housing problem', *Housing Studies* 3(3): 190–5.

Economic Planning Agency, Government of Japan (1988) *Annual Report on the*

*National Life for FY 1987: The Higher Yen and Forming Affluent Infrastructures.*

Elkinton, J. and Burke, T. (1989) *The Green Capitalists*, Victor Gollancz, London.

Environment Agency, Planning Committee of the Central Council for Environmental Pollution Control (1972) *Long-Term Prospects for Preservation of the Environment: An Interim Report*, Tokyo.

Environment Agency (Sep. 1974) 'Guidelines for conducting environmental impact assessment (interim report)', *Japan Environment Summary* 2(9): 1–4.

—— (Feb. 1975) 'Experts report on a legal system for environmental impact assessment', *Japan Environment Summary* 4(2).

—— (Aug. 1977) *Japan Environment Summary* 5(8).

—— (Aug. 1978) *Japan Environment Summary* 6(8).

—— (Oct. 1978) 'Survey of local government environmental protection measures', *Japan Environment Summary* 16(2).

—— Central Council on Environmental Pollution Control (May 1979) 'Council recommendations on a system of environmental impact assessment', *Japan Environment Summary* 7(5).

—— (Nov. 1979) 'Recent trends in establishing EIA procedures by local government', *Japan Environment Summary* 7(11).

—— Central Council for Environmental Pollution Control (Mar. 1981) 'Environmental policy in the 1980s: tasks and directions', *Japan Environment Summary* 9(3).

—— (Nov. 1982, unpublished, in Japanese) 'The environmental impact assessment bill: proceedings of the 92nd Diet session'.

—— (Nov. 1984) 'Cabinet decision on the implementation of environmental impact assessment', *Japan Environment Summary* 12(11).

—— (Feb. 1985) 'Principles for implementing environmental impact assessment', *Japan Environment Summary* 13(2).

—— Coordination Bureau (Mar. 1985, in Japanese) *Environmental Impact Assessment Technical Investigation*.

—— (Apr. 1986) 'Regional environmental management planning guidebook completed', *Japan Environment Summary* 14(4).

—— (1986) *Introduction to the Environment Agency of Japan*.

—— Coordination Bureau (June 1987, in Japanese) *Ministry Guidelines for Impact Assessment*, Tokyo.

—— (Mar. 1987, unpublished) 'Environmental impact investigations support systems: EIA system case study material report'.

—— (July 1987) Management and Coordination Bureau, 'Concerning the environmental impact assessment for the Trans-Tokyo Bay Highway, press release.

—— (various years) *Quality of the Environment in Japan*.

*Environment Assessment Handbook* (1988, in Japanese) Musashino Publishers, Tokyo.

Foreign Press Centre (Oct. 1987) *Outline of the FY 1986 White Paper on National Land Use*, W-87-6, Tokyo.

—— (1987) *Fact and Figures of Japan*, Tokyo.

Forrest, R.A. (Dec. 1986) *Kogai to Gaiko: Japan and the World Environment*, Center for Japan Studies, Horace Rackham School of Graduate Studies, University of Michigan, Ann Arbor.

Fujita, K. (1988) 'The technopolis: high technology and regional development in Japan', *International Journal of Urban and Regional Research* 12(4): 566–93.

General Agreement on Tariffs and Trade (1986) *International Trade 1985–6*, Geneva.

Glasmeier, A.K. (1988) 'The Japanese technopolis programme', *International Journal of Urban and Regional Research* 12(2): 268–83.

Glasson, J. and Elson, M. (Aug. 1987) 'The planning and inquiry process and infrastructure projects', Paper No. 3 presented at the Fifth Annual Conference of the Major Projects Association, Oxford.

Gottman, J. (1980) 'Planning and metamorphosis in Japan: a note', *Town Planning Review* 51(2): 171–6.

Gresser, J., Fujikura, K., and Morishima, A. (1981) *Environmental Law in Japan*, The MIT Press, Cambridge, Massachusetts.

Hase, T. (1981a) 'Japan's growing environmental movement', *Environment* 23(2): 14–36.

—— (1981b) 'The Japanese experience', in O'Riordan and Sewell (1981): 227–51.

Hashimoto, M. (1985) 'Development of environmental policy and its institutional mechanisms of administration and finance', presented at the International Workshop on Environmental Management for Local and Regional Development, UN Environment Programme, 9–13 June, Nagoya.

Hashimoto, Z. (1985) *Information System for Environmental Management for Local and Regional Development*, UN Environment Programme, Nagoya.

Hattori, H. (1987) 'Environment management program in Nagoya', presented at the First Pacific Environment Conference, 15–20 June, Naggoya.

Hebbert, M. (1986) 'Urban sprawl and urban planning in Japan' *Town Planning Review* 57(2): 141–58.

Hiraiwa, T. (1979) *The New Kansai International Airport Document*, Akishobo Publishers, Tokyo.

Hiraoka, M. (Mar. 1986) 'Solid waste management', presented to the International Course for Graduate Research Students in the Field of Civil Engineering, Kyoto University, Kyoto.

Honshu–Shikoku Bridge Authority (1986) *Outline of Honshu–Shikoku Bridge Project*, Tokyo.

Huddle, N. and Reich, M. (1975) *Islands of Dreams: Environmental Crisis in Japan*, Autumn Press, Japan.

Industrial Pollution Control Association of Japan (1983) *Environmental Protection in the Industrial Sector in Japan*, Tokyo.

International Cultural Association (1990) *Who's Who in Japanese Government 1990/91*.

International Society for Educational Information Inc. (1987) *Understanding Japan: Japan's Industrial Economy – Recent Trends and Changing Aspects*, Shoobi Publishing Co., Tokyo.

International Union for the Conservation of Nature, Species Survival Commission (Mar. 1988) *Shiraho Coral Reef and the Proposed New Ishigaki Airport, Japan: Preliminary Report*, prepared by the International Marine Alliance Canada, Ottawa.

—— (July 1988) *Review of the Draft Environmental Assessment of the Landfill Operations Attendant on the Construction of the New Ishigaki Airport, Okinawa Prefecture, Japan, April 1988*.

Isobe, C., Kato, I., Namura, Y., Morishima, A. and Yamamura, T. (1979, in Japanese) 'Legal changes associated with environmental assessment', *Jurist* 695: 15–30.

Iwata, M. (1986, unpublished) 'Environmental impact assessment of the KIA', Environment Agency, Tokyo.

Japan Electric Power Information Centre (1986) *Nuclear Power in Japan*, Tokyo.

Japan Electricity Production Information Centre (1988) *Electric Power Industry in Japan 1988*, Tokyo.

Japan External Trade Organization (1983) *Environmental Control in Japan*, marketing series 15.
—— (various years) *Business Facts and Figures*.
Japan Highway Public Corporation (1987) *Environment Assessment for the Trans-Tokyo Bay Highway*.
Japan Transport Economics Research Centre (1982) *Transportation in Japan*, Tokyo.
Kajima Construction Co., Environmental Development Div. (1987, in Japanese) *Environmental Impact Assessment in Practice*, Kajima Publishing Co., Tokyo.
Kansai International Airport Co. (Oct. 1985, in Japanese) *Draft Environmental Impact Statement for the New KIA Development*, Osaka.
—— (1986) *Introducing the Kansai International Airport*, Osaka.
—— (June 1986, in Japanese) *Environmental Impact Statement for the New Kansai International Airport Development*, Osaka.
—— (Feb. 1987) *Introducing the Kansai International Airport*, Osaka.
—— (May 1987) *Understanding More About the KIA*, Osaka.
Kawashima, T. and Stoehr, W. (1988) 'Decentralized technology policy: the case of Japan', *Environment and Planning C: Government and Policy* 6: 427–39.
Keizai Koho Centre, Japan Institute for Social and Economic Affairs (1988) *Japan 1989: An International Comparison*, Tokyo.
Keizai Shunjusha Co. Ltd (1972, in Japanese) *The Pollution Handbook*, Tokyo.
Kelley, D.R., Stunkel, K.R. and Wescott, R.R. (1976) *The Economic Superpowers and the Environment: The US, the Soviet Union and Japan*, W.H. Freeman & Co., San Francisco.
Kihara, K. (1981) 'Japan's environmental policies: the last ten years', *Japan Quarterly*: 501–8.
Kiyoaki, T. (1987) *Public Administration in Japan*, University of Tokyo Press, Tokyo.
*Kodansha Encyclopaedia of Japan* (1985), Kodansha Publishing Co., Tokyo.
Koller, J.M. (1985) *Oriental Philosophies*, Charles Scribner's Sons, New York.
Kondo, S.(1981) 'Summary procedures for the settlement of pollution cases', in Kato, I. *et al.*, (eds), *Environmental Law and Policy in the Pacific Basin Area*, University of Tokyo Press, Tokyo.
Kunamoto, N. (1981) 'Recent tendencies and problems of court cases on environmental protection in Japan', in Kato, I. *et al.*, (eds), *Environmental Law and Policy in the Pacific Basin Area*, University of Tokyo Press, Tokyo.
Kurihara, S., Kubokawa, T. and Nakashima, S.R. (1982) 'Japan's EIA procedure for nuclear facilities', *EIA Review* 3(2–3): 289–95.
Kyoto Prefecture (1989) Guidelines for Environmental Impact Assessment.
McKean, M.A. (1981) *Environmental Protest and Citizen Politics in Japan*, University of California Press, Berkeley.
Masser, I. (1985) 'Japanese urban planning: some British perspectives', *Town Planning Review* 57(2): 123–6.
Ministry of Foreign Affairs (1973/4) *Development of Environmental Protection in Japan*, Tokyo.
Ministry of International Trade and Industry (Dec. 1988) *Energy in Japan: Facts and Figures*, Tokyo.
Ministry of Transport, Aviation Deliberation Council (Aug. 1974) *Size and Location of the Kansai International Airport: Findings*, Tokyo.
Ministry of Transport (1981a, in Japanese) *Draft Environmental Impact Assessment for the KIA*, Tokyo.
—— (1981b, in Japanese) *Plan for the Development of the Kansai International Airport-Related Area*, Tokyo.

—— (1985) *Urban Transportation in Japan*, Tokyo.

—— (1988) *Annual Report on the Transport Economy: Summary (Fiscal 1987)*, Tokyo.

Mitchell, B.R.(1982) *International Historical Statistics: Africa and Asia*, Macmillan Press, London.

Moore, C.A. (ed.) (1967) *The Japanese Mind*, East–West Center Press, Honolulu.

Mori, Y. (Aug. 1989) 'Some issues on coordination of government policy and local ordinances on EIA', *Environmental Systems Research* 17: 195–201.

Morishima, A. (1981) 'Japanese environmental policy and law', in Kato, I. *et al.* (eds) *Environmental Law and Policy in the Pacific Basin Area*, University of Tokyo Press, Tokyo.

Morita, T. (1981) *Policy for Plan Assessment: Present Research Study*, National Institute for Environmental Studies, Tsukuba.

—— and Gotoh, S.(1985) 'A study of the effects of implementing environmental impact assessment procedure', Research Report No. 81, National Institute for Environmental Studies, Tsukuba.

Munn, F.E. (ed.) (1979) *Environmental Impact Assessment*, John Wiley, Chichester.

Muzik, K. (1985) *Dying Coral Reefs of the Ryukyu Archipelago (Japan)*, Proceedings of the Fifth International Coral Reef Congress, Vol. 6, Tahiti: 483–9.

Nakai, N. (1988) 'Urbanization promotion and control in metropolitan Japan', *Planning Perspectives* 3: 197–216.

Nakamura, H. (1964) *Ways of Thinking of Eastern Peoples*, The University Press of Hawaii, Honolulu.

—— (Sept. 1989) 'Kankyo Hyaka: zoom-up', Shin Nihon Publishers, Tokyo.

Nakane, C. (1986) *Japanese Society*, Charles E. Tuttle Co., Tokyo.

Namiki, O. (July–Sep. 1985) 'The unhappy birth of a tunnel', *Japan Quarterly* 32(3): 324–9.

Nomura, Y., Asano, N., Ogano, S. and Inoue, H. (1985) 'Roles of environmental control plans, EIA reports and environmental information in Japan today', UNEP International Workshop on Environmental Management for Local and Regional Development, 9–13 June, Nagoya.

—— (1987) 'Roles of environmental control plans, environmental impact assessment report and environmental information: situation in Japan today', presented at the First Pacific Environment Conference, 15–20 June, Nagoya.

OECD (1977) *Environmental Policies in Japan*, Paris.

Okita, S. (1989) 'Saving our environment', *Journal of Japanese Trade and Industry* 5: 8–11.

Okinawa Prefectural Government (Apr. 1988, in Japanese) *Draft Assessment Report for the Reclamation Works for the New Ishigaki Airport: Summary*, Okinawa.

O'Riordan, T. and Hey, R. (eds) (1976) *Environmental Impact Assessment*, Saxon House, Aldershot.

—— and Sewell, W.R.D. (eds) (1981) *Project Appraisal and Policy Review*, John Wiley & Sons, Chichester, England.

Osaka Lawyers' Shiraho Study Group (Mar. 1989) 'Coral for Future Generations: a statement in favour of reassessing the New Ishigaki Airport Plan', Osaka.

Osaka Prefectural Government, Planning Division (1983, in Japanese, unpublished) 'Kansai International Airport related materials', Osaka.

—— Living Environment Division (Apr. 1984) *Osaka Prefecture Environmental Impact Assessment*, Osaka.

—— (1986a, in Japanese, unpublished) 'Record of public hearings concerning the Kansai International Airport-related developments and South Osaka coastal developments', Osaka.

—— (1986b, unpublished, in Japanese) 'Regional development and the Kansai International Airport', Osaka.

—— (June 1986, in Japanese) *Environmental Impact Statement for South Osaka Coastal Developments*, Osaka.

Osaka Prefecture (1987) 'The creation of a new international "aeropolis" in complete harmony with the Kansai International Airport', Osaka.

Oshima, T. (1989) 'Environmental impact assessment', in *Kyoto University Civil Engineering Handbook 1988–9*, Kyoto.

Prime Minister's Information Office, Public Opinion Inquiry Division (Aug. 1987, in Japanese) *Public Opinion Inquiry Concerning Nuclear Power*, Tokyo.

Reich, M. (1983) 'Environmental policy and Japanese society: part 1 successes and failures' *International Journal for Environmental Studies* 20: 191–8.

Reischauer, E.O. (1986) *The Japanese*, Charles E. Tuttle Co., Tokyo.

Sakurai, M. (1988) 'Environmental impact assessment system in Japan', presented at the Expert Group Meeting on Environmental Impact Assessment of Development Projects, 15–19 Aug., Bangkok.

Saruta, K. (1984, in Japanese) 'Trends in local laws', *Environment* 9(2): 14–29.

Shimazu, Y. (1987) *Environmental Impact Assessment*, NHk Books, Japan.

—— and Harashima, R. (1977) 'Environmental assessment–management system for local development project: cases in Japan', *Journal of Environmental Management* 5: 243–58.

Shiraho Protection Group (Apr. 1988) *Let the Sea Stain Our Hearts*, Okinawa.

Short, K. (1989) 'Tokyo bay: crabs and concrete – an overview of changes on the Tokyo Bay ecosystem', *Japan Environment Monitor* 2/3: 1, 6–7, 16.

Siman, B.B. (Mar. 1989) 'Land use planning in Japan', *The Planner*: 13–16.

Stewart-Smith, J. (1987) *In the Shadow of Fujisan*, Penguin Books, Middlesex, England.

Suzuki, M. (spring 1986) 'Battle for Shiraho coral reef', *Japan Environment Review*: 10–18.

—— (Dec. 1986) 'Report on the Shiraho reef/New Ishigaki Island Airport Controversy', Friends of the Earth, 15–20 June, Nagoya.

—— (Aug. 1988) 'Impact of the IUCN resolution: publication and criticism of the revised environmental assessment draft report', *Shiraho News*.

—— (Sep. 1989) 'Biological diversity and the conservation of natural habitats – the role of Japan', briefing paper for International People's Forum on Japan and the Global Environment, 8–10 Sep., Tokyo.

—— (various dates) 'Shiraho News', *Japan Environment Monitor*, Tokyo.

Suzuki, Y. (1985, in Japanese) 'The Cabinet decision on environmental impact assessment', *Environmental Research* 53: 52–62.

Takada, T. (1978) 'Noise pollution on trial: the case of Osaka International Airport', *Japan Quarterly* 25(1): 35–42.

Tamura, A. (Oct.–Dec. 1987) 'Deconcentrating Tokyo, reconfiguring Japan', *Japan Quarterly*: 35–42.

Therivel, R. (1988) *Environmental Impact Procedures in Japan*, Kyoto University.

Tokyo Metropolitan Government (1971) *Pollution Control in Tokyo*, Tokyo.

Trade and Industry Technological Data Research Co. (1975, in Japanese) *Industrial Pollution and Administration*, Tokyo.

Trans-Tokyo Bay Highway Co. (May 1987) *The Trans-Tokyo Bay Highway Development*, Tokyo.

Tsuji, K. (1984) *Public Administration in Japan*, University of Tokyo Press, Tokyo.

Tsukatani, T. (Sep. 1987) 'Current prospects for the Japanese compensation of pollution related health damage', Discussion Paper No. 237, Kyoto Institute of Economic Research, Kyoto University, Kyoto.

—— (1989) 'Compensation system for environmental damage and other economic incentive in Japan', Discussion Paper No. 275, Kyoto Institute of Economic Research, Kyoto University, Kyoto.

Twijnstra Gudde NV Management Consultants (Feb. 1979) *Summary Report: Recommendations on EIA on the Basis of Trial Runs and Complementary Research*, Ministry of Health and Environmental Protection, Dutch Government.

Uchino, T. (1978) *Japan's Postwar Economy*, Kodansha International, Tokyo.

Van Wolferen, K. (1989) *The Enigma of Japanese Power – People and Politics in a Stateless Nation*, MacMillan, London.

Weidner, H. (1986) 'Japan: the success and limitations of technocratic environmental policy', *Policy and Politics* 8(1): 43–70.

Westman, W.E. (1985) *Ecology, Impact Assessment, and Environmental Planning*, John Wiley & Sons, New York.

Yamaguchi, S. (Aug. 1984) 'Japan: towards a new metropolitan policy', *Cities*: 474–86.

Yamamura, K. and Yasuba, Y. (1987) *The Political Economy of Japan*, Stanford University Press, Stanford.

Zetter, J. (1986) 'Challenges for Japanese urban policy', *Town Planning Review* 57(2): 135–40.

## Interviews

Hase, T., Public Health Department, Kyoto Municipal Government, 1987–8.

Iwata, I., Environmental Impact Assessment Division, Environment Agency, 27 July 1987.

Kariya, Y., Planning Bureau, Kyoto Municipal Government, 26 Feb. 1988.

Kato, H., Consultant, Kansai Environmental Engineering Centre, 27 Feb. 1988.

Kawada, K., Corporate Planning Department, Kansai International Airport Company Ltd, 20 July 1987.

Miki, Y., Japan Green Party, Osaka, 10 Oct. 1987.

Morita, T., Systems Analysis and Planning Division, National Institute for Environmental Studies, Tsukuba, 3–4 Mar. 1988.

Nakamura, K., Social Sciences Department, Tsukuba University, 4 Feb. 1988.

Rockow, J., University of Pittsburgh, Ritsumeikan University, Kyoto, 29 Oct. 1988.

Shimada, Y., Treasury Division, Kansai International Airport Co., 20 July 1987.

Sumitomo, H., Professor, Department of Environmental Engineering, Kyoto University, 1987–8.

Suzuki, M., Environmentalist, Friends of the Earth (Japan), 1987–8.

Suzuki, T., National Institute for Public Health, 8 Aug. 1989.

Takatsuki, H., Environmental Conservation Centre, Kyoto University, 1988–9.

Takeda, F., Highway Research Group, 9 Aug. 1989.

Toda, T., Civil Engineering Department, Kyoto University, 1988.

Tsukatani, T., Economic Research Institute, Kyoto University, 1988–9.

Yanaka, S., Environmentalist, Ishigaki Island, 5 Aug. 1988.

Yamasato, S., Citizens' Group Spokesperson, Shiraho Village, Ishigaki Island, 24 July and 5 Aug. 1988.

# Index